P. Riederer, D. B. Calne,
R. Horowski, Y. Mizuno, W. Poewe,
M. B. H. Youdim (eds.)

Advances in Research
on Neurodegeneration

Volume 5

Springer-Verlag Wien GmbH

Prof. Dr. P. Riederer

Department of Psychiatry, University of Würzburg, Federal Republic of Germany

Prof. Dr. D. B. Calne

Neurodegenerative Disorders Centre, University of British Columbia, Vancouver, Canada

Dr. R. Horowski

Clinical Research, Schering AG, Berlin, Federal Republic of Germany

Prof. Dr. Y. Mizuno

Department of Neurology, Juntendo University, Tokyo, Japan

Prof. Dr. W. Poewe

Department of Neurology, University of Innsbruck, Austria

Prof. Dr. M. B. H. Youdim

Institute of Pharmacology, Technion – Israel Institute of Technology, Faculty of Medicine, Haifa, Israel

© 1997 Springer-Verlag Wien
Originally published by Springer-Verlag Wien New York in 1997
Softcover reprint of the hardcover 1st edition 1997

Printing: A. Holzhausens Nfg., A-1070 Wien
Graphic design: Ecke Bonk
Printed on acid-free and chlorine-free bleached paper

With 45 Figures

ISBN 978-3-211-82898-4   ISBN 978-3-7091-6842-4 (eBook)
DOI 10.1007/978-3-7091-6842-4

Preface

Volume 5 of the series "Advances in Research on Neurodegeneration" is concerned with themes which are currently the focus of intensive research, and in which advances in our understanding of the pathological mechanisms underlying neurodegenerative diseases are expected in the near future. The first section contains five reviews devoted to the various neuroimaging technologies. The discussion is concerned with the question of whether neuroimaging techniques make it possible to follow the process of degeneration as it occurs, and which methods offer the required sensitivity and quantifiability for this purpose. However, the question needs to be examined of whether, given the physical and chemical limitations of these techniques, even under optimal conditions, anatomical resolution can be improved to the extent that neurodegenerative diseases can be diagnosed earlier than currently possible and a confident diagnosis made. The possibilities of using neuroimaging techniques to provide information regarding the effects of neuroprotective or neuroregenerative therapeutic strategies, and for correlating the results of neuropsychological research with imaging data are also discussed.

The second section is concerned with the significance of endogenous or exogenous neurotoxins as triggers for neurodegenerative processes that may lead to Parkinsonism. Vulnerability factors, which include such factors as nerve ending sensitivity, the synergistic effects of drugs and the various mechanisms underlying different toxins are discussed.

A further important chapter considers the question of cell death; that is, programmed cell death, apoptosis and necrosis, and their respective mechanisms. Questions regarding terminology, as well as molecular biological and genetic mechanisms, are the main points of interest. The validation and limitations of neuropathological research methods, as well as possible therapeutic consequences are also discussed.

The fourth chapter is concerned with new developments in the area of immunoinflammatory mechanisms and infectious diseases which may trigger neurological diseases. Both basic experimental studies and clinically oriented research are reviewed. The focus is on multiple sclerosis and prion diseases.

The rapid publication of this volume provides the reader with an up-to-date overview of these themes in a compact form.

The publication of the Proceedings of the Fifth International Winter Conference on Neurodegeneration (Spitzingsee, Bavaria, February 1996) was made possible by the generous assistance of Schering AG, Berlin, and Lilly, Bad Homburg, Germany. We thank Springer-Verlag Wien New York for their usual excellent quality and swiftness.

Würzburg, December 1996   **P. Riederer**

Contents

Brain imaging revisited

Endogenous and exogenous neurotoxins

Programmed cell death, apoptosis, necrosis and in between

Immunoinflammatory mechanisms, infective diseases causing neurological disorders

Listed in Current Contents / Life Sciences

Brain imaging revisited

Chair: W. Poewe and D. B. Calne

Magnetic Resonance: A multimodal approach to the brain?

J. Laubenberger

University Clinic of Radiology, Freiburg i. Br., Federal Republic of Germany

Summary. Magnetic resonance of the brain offers the unique possibility of acquisition of morphological data in high resolution quality, analysis of metabolic disturbances, and the assessment of brain function. Basic principle is the BOLD effect, using the difference of the magnetic moments of oxy- and deoxyhemoglobin as intrinsic mechanism of contrast. Still the optimum techniques for functional imaging are not established: echo planar imaging requires a very costly hardware, whereas gradient echo imaging is feasible on clinical routine MR scanners. Although many experimental studies have been published on visual, motor, and auditory paradigms, clinical applications have not been established up to now. Analysis of epileptogenic foci combining high resolution imaging and functional information seems very promising as well as preoperative determination of critical areas in neurosurgery.

Introduction

MR imaging of the brain has opened new horizons in diagnosing neurological diseases. In addition to the morphological description of the brain tissue by conventional imaging, and the depiction of metabolic abnormalities by MR spectroscopy, assessment of brain function has become feasible.

In 1990 Belliveau published the first observation of the stimulation of the human visual cortex by magnetic resonance imaging (Belliveau et al., 1990, 1991). He used the observation of the first pass effect after bolus injection of a contrast agent to demonstrate changes in cortical perfusion upon activation with a photic stimulus. The use of bolus tracking to study changes in perfusion was an exact analogue to previous experiments using the observation of tracers with PET or SPECT. It required the injection of contrast agent in two consecutive scans, one with and one without stimulus. The performance of such an experiment with MR instead of nuclear medicine techniques offers the immediate advantage of vastly superior spatial and temporal resolution and the lack of radioactive tracers. The need for dual contrast agent injection nevertheless poses a problem especially for studies of brain activation in normals. This disadvantage was resolved by the demonstration of brain activation using the BOLD-contrast mechanism first described by Ogawa (Ogawa et al., 1990). This elegant technique has led to a fast proliferation of functional MRI (fMRI) in various centers over the last few years.

The BOLD-contrast

The BOLD-contrast relies on the fact, that deoxyhemoglobin – in contrast to oxyhemoglobin – has a strong magnetic moment (Thulborn et al., 1982). By interaction of the bulk magnetization of deoxygenized blood with the external field local field variations in and around blood vessels are thus created. These so-called susceptibility effects can be measured using appropriate MR imaging sequences.

The only source of energy of normal brain cells is the oxidation of glucose. Since the glucose storage capacity of brain cells is negligible, the brain very heavily depends on a constant supply of glucose and oxygen via the capillary bed. If the cortex is activated, its oxygen (and glucose) demand increases. This increased demand leads to a fast and local dilation of the capillaries via a mechanism, which is not fully understood (Fox et al., 1986). The amount of blood flowing to the activated area is thus increased. The additional blood supply in most known instances overcompensates the increased oxygen consumption. The local concentration of oxygenated blood is thus increased. This leads to a decrease in the local susceptibility effect, which can be visualized using fMRI.

Experimental techniques

The first brain activation studies by Kwong used gradient echo planar imaging (GR-EPI) (Kwong et al., 1992). The EPI-sequence uses multiple gradient refocusings in order to acquire all data necessary for image reconstruction after a single excitation pulse. Each echo is thus read out at a different echo time. The overall image contrast is given by the echo time of the projections carrying low phase encoding, which for typical such experiments lies in the order of 30–50 ms. Since each projection experiences different susceptibility weighting it should be noted, that stimulation does not only changes the signal intensity but also the point spread function of the voxel under observation. In spite of this not very well defined signal behavior, EPI has turned out to be a very efficient technique for brain activation studies due to its short acquisition time.

Conventional gradient echo imaging with long echo times (40–60 ms depending on field strength) has also turned out to be a useful technique for fMRI (Frahm et al., 1992, 1993). Its advantage over EPI lies in the fact, that it allows the acquisition of high resolution images, whereas the resolution in EPI is determined roughly by the number of echoes which can be acquired within the T2 of brain parenchyma, which is in the order of 60 ms. Although a higher nominal resolution can be achieved using longer acquisition times, the actual resolution will be limited by blurring caused by T2*-decay. It should be noted, that the apparent sharpness of EPI-images of the brain, which are acquired with acquisition times of 100–200 ms is afforded by the high resolution of the CSF-filled spaces due to the long T2 of CSF.

A further practical advantage of gradient echo techniques is their much wider availability on clinical scanners. fMRI with gradient echoes is widely abundant, whereas EPI still is a rare and costly commodity. Conventional gradient echo imaging does, however, suffer from a number of severe drawbacks. The long acquisition time per image restricts the application to a single slice and thus requires prior knowledge about the area of activation. Partial volume effects can lead to difficulties in data interpretation. Gradient echo techniques also are very sensitive to inflow (Hajnal et al., 1993; Duyn et al., 1993). Since vascular flow in vessels – especially large veins – also changes upon stimulation, this can lead to the measurement of activation effects many centimeters from the area of activation (Segebarth et al., 1994). These vascular signal changes can be much larger than the actual parenchymal effects, which very seldom exceed 2–3%. The use of low flip-angles ameliorates this problem and leads to less pronounced vessel effects. A slightly more demanding procedure to avoid inflow effects at least in large vessels is the use of saturation slices parallel and immediately adjacent to the slice under study. As a consequence of the inflow problem, results from gradient echo measurements performed with flip angles larger than 25–30° and/or effects larger than 5% should be distrusted.

Finally gradient echo techniques use a susceptibility sensitizing time interval for each projection step. During each such time interval effects like involuntary motion or system instabilities will introduce changes in the signal phase. The stochastic nature of these effects will lead to artifacts in the final image. Such artifacts can be reduced at least in principle using navigator echo techniques (Kim et al., 1990) or more elaborate matching algorithms (Ehman et al., 1989; Hajnal et al., 1994), which are, however, not available on clinical scanners.

The current trend in fMRI-methodology is the transition from two-dimensional (2D-) measurements to three-dimensional (3D-) data. These offer some highly significant advantages over 2D-images. Areas of possible stimulation must not be sought painstakingly before the examination and partial volume effects are avoided. This allows the acquisition of much more reproducible data for the examination of long term stability of activation patterns. Distributed stimulation in different areas can be studied. These advantages outweigh in most instances the disadvantages of longer acquisition times compared to 2D-experiments. The total acquisition time should nevertheless not exceed 30–40 s in order to ensure constant activation during the stimulation periods. With conventional gradient echo imaging such volumetric examinations are not feasible due to the long acquisition times. Volumetric examinations with EPI are currently be performed mostly as multislice experiments in order to avoid saturation and instability problems. Other techniques, which are currently under development for that purpose are modified RARE-sequences (Hennig et al., 1994) and the MUSIC-technique (Loenneker et al., 1994).

Finally functional spectroscopy (Hennig et al., 1994) has to be mentioned as a method not for localizing centers of activation but to study the temporal evolution of activation at a predefined localization. Compared to imaging methods it combines a high effect-to-noise ratio and high temporal resolution at the cost of low spatial resolution. Apart from various neurophysiological studies it has been applied to demonstrate a short-term effect immediately after stimulation (Ernst et al., 1994).

Applications

The first experiments performed with fMRI used the well-known paradigm of photic stimulation with a flicker display or an alternating checkerboard pattern. This is known to lead to significant changes in perfusion and thus serves as a test tool for sequence development. Meanwhile quite a number of experiments have been performed, which led to new insights in neurocognitive research. Apart from activation in the primary visual cortex, activation of associated areas was demonstrated using a number of paradigms to test cognitive processing of motion, texture, color, object recognition and others (First Annual Meeting SMR, 1994). Various paradigms using motor activation has been successfully examined (Bandettini et al., 1992; Kim et al., 1993). Numerous groups have investigated language processing using a number of well established paradigms (Benson et al., 1993, 1994; Hinke et al., 1993). Apart from activation of the cerebral cortex, the involvement of the cerebellum in learning tasks has been demonstrated (Ellermann et al., 1994). Subcortical activation has been found for example in the geniculate nucleus (upon visual stimulation) (Frahm et al., 1993). In view of the already considerable body of knowledge produced with fMRI it is safe to assume, that MR will be the main technique for localizing centers of activation in neurocognitive research and will replace more conventional modalities like PET and SPECT for these applications. The main advantages are the reduced cost and reduced problems with irradiation exposure on one hand. Even more important is the possibility of fMRI to perform repetitive examinations with multiple repetitions and thus to significantly improve the significance of the results. Interindividual comparisons, which are extremely difficult with PET, can be easily studied with fMRI.

A number of clinical applications have been proposed with fMRI. Lateralisation of epileptic foci has been demonstrated (Gadian et al., 1994). The very high sensitivity of the fMRI-examination to motion makes it, however, doubtful, whether this application will be significant in the future especially since other more robust techniques like MR spectroscopy exist for that purpose.

A second clinical application would be the localization of sensitive areas in neurosurgical operation planning. fMRI has been demonstrated to be applicable for that purpose and to yield comparable results with PET (Hertz-Pannler et al., 1994; Howard et al., 1994). The problem is of course, that an exhaustive search for all possible high priority areas is not possible due to the lack of effecient paradigms. On the other hand, pharmakological testing during surgery is possible in most patients, so there is some uncertainty, whether there is a real demand for this type of application.

Conclusions

fMRI has in its short history already proven to be a highly relevant tool in neurocognitive research, and a part of a multimodal diagnostic approach to the brain. However, direct clinical applications are scarce and under discussion. The main problem today is the lacking availability of robust techniques on clinical scanners expressed most notably by the multitude of examinations using gradient echo techniques with inadequate measurement parameters. The specifications regarding long term stability of the systems required for fMRI are of no concern for conventional MR diagnosis and are thus frequently not being met even by state-of-the art scanners.

References

Bandettini PA, Wong EC, Hinks RS, Tikofsky RS, Hyde JS (1992) Time course EPI of human brain during task activation. Magn Res Med 25: 390–397

Belliveau JW, Kennedy DN, McKinstry RC, Buchbinder BR, Weisskoff RM, Cohen MS, Vevea JM, Brady TJ, Rosen BR (1991) Functional mapping of the human visual cortex by magnetic resonance imaging. Science 254: 716–717

Belliveau JW, Rosen BR, Kantor HL, Rzedzian RR, Kennedy DN, McKinstry RC, Vevea JM, Cohen MS, Pykett IL, Brady TJ (1990) Functional cerebral imaging by susceptibility-contrast NMR. Magn Res Med 14: 538–546

Benson RR, Kwong KK, Belliveau JW, Baker JR, Cohen MS, Hildebrandt N, Rosen BRT (1993) Proc XIIth Ann Meeting SMRM, New York, p 1398

Benson RR, Kwong KK, Buchbinder BR, Jiang HJ, Belliveau JW, Cohen MS, Bookheimer S, Rosen BR, Brady T (1994) Proc Ist Ann Meeting SMR, San Francisco, p 684

Duyn JH, Moonen CTW, de Boer RW, Yperen GH, Luyten PR (1993) Inflow versus deoxyhemoglobin effects in "BOLD" functional MRI using gradient echoes at 1.5T. Proc XIIth Ann Meeting SMRM, New York, p 168

Ehman RL, Felmlee JP (1989) Adaptive technique for high definition MR imaging of moving structures. Radiology 173: 255–263

Ellermann JM, Flament D, Kim SG, Merkle H, Andersen P, Ebner T, Ugurbil K (1994) Cerebellar activation due to error detection-correction in a visuo-motor learning task: a functional magnetic resonance study. Proc Ist Ann Meeting SMR, San Francisco, p 331

Ernst T, Hennig J (1994) Observation of a fast response in functional MR. Magn Reson Med 32: 146–149

First Ann Meeting SMR, San Francisco (1994) Proceedings, pp 333, 671, 672, 677, 697, 688–692, 695, 700, 704

Fox PT, Raichle ME (1986) Focal physiological uncoupling of cerebral blood flow and oxidative metabolism during somatosensory stimulation in human subjects. Proc Natl Acad Sci USA 83: 1140–1144

Frahm J, Bruhn H, Merbold KD, Hanicke W (1992) Dynamic MR imaging of the human brain oxygenation during rest and photic stimulation. J Magn Reson Imag 2: 501–505

Frahm J, Merboldt KD, Hänicke W (1993) Functional MRI of human brain activation at high spatial resolution. Magn Reson Med 29: 139–144

Frahm J, Merboldt KD, Hänicke W, Kleinschmidt A, Steinmetz H (1993) High-resolution functional MRI of focal subcortical activity in the human brain. Long-echo time FLASH of the lateral geniculate nucleus during visual stimulation. Proc XIIth Ann Meeting SMRM, New York, p 57

Gadian DG, Cross JH, Gordon I, Jackson GD, Neville BGR, Connelly A (1994) 1H MRS and interictal spect in children with intractable temporal lobe epilepsy. Proc Ist Ann Meeting SMR, San Francisco, p 344

Hajnal JV, Myers R, Oatridge A, Schwieso JE, Young IR, Bydder GM (1994) Artifacts due to stimulus correlated motin in functional imaging of the brain. Magn Res Med 31: 283–291

Hajnal JV, Oatridge A, Schwieso J, Cowan FM, Young IR, Bydder GM (1993) Cautionary remarks on the role of veins in the variability of functional imaging experiments. Proc XIIth Ann Meeting SMRM, NewYork, p 166

Hennig J, Ernst Th, Speck O, Deuschl G, Feifel E (1994) Detection of Brain activation using oxygenation sensitive functional spectroscopy. Magn Res Med 31: 85–90

Hennig J, Hennel F, Oesterle C, Speck O, Janz C, Nedelec JF (1994) Fast and robust measurements of brain activation using modified RARE-sequence with variable contrast. Proceedings of the Society of Magnetic Resonance, Second Meeting, San Francisco, p 660

Hertz-Pannler L, Gaillard WD, Mott S, Cuenod CA, Bookheimer S, Weinstein S, Conry J, Theodore WH, Le Bihan D (1994) Pre-operative assessment of language lateralization by FMRI in children with complex partial seizures: Preliminary study. Proceedings of the Society of Magnetic Resonance, Second Meeting, San Francisco, p 326

Hinke RM, Hu X, Stillman AE, Ugurbil K (1993) The use of multislice functional MRI during internal speech to demonstrate the lateralization of language function. Proc XIIth Ann Meeting SMRM, New York, p 63

Howard R, Alsop D, Detre J, Listerud J, Zager E, Judy K, Flamm E, Hurst R, Atlas SW (1994) Functional MRI of regional brain activity in patients with intracerebral gliomas and AVMs prior to surgical or endovascular therapy. Proceedings of the Society of Magnetic Resonance, Second Meeting, San Francisco, p 701

Kim SG, Ashe J, Hendrich K, Ellermann JM, Merkle H, Ugurbil K, Georgopoulos AP (1993) Functional magnetic resonance imaging of motor cortex: Hemispheric asymmetry and handedness. Science 261: 615–617

Kim WS, Mun CW, Kim DJ, Cho ZH (1990) Extraction of cardiac and respiratory motion cycles by use of projection data and its application to NMR imaging. Magn Reson Med 13: 25–37

Kwong KK, Belliveau JW, Chesler DA, Goldberg IE, Weisskoff RM, Poncelet BP, Kennedy DN, Hoppel BE, Cohen MS, Turner R, Cheng HM, Brady TJ, Rosen BR (1992) Dynamic magnetic resonance imaging of human brain activity during primary sensory stimulation. Proc Natl Acad Sci USA 89: 5675–5679

Loenneker T, Hennig J (1994) MUSIC. A fast T2*-sensitive MRI technique with enhanced volume coverage. Proc Natl Acad Sci India (in press)

Ogawa S, Lee TM, Nayak AS, Glynn P (1990) Oxygenation-sensitive contrast in magnetic resonance image of rodent brain at high magnetic fields. Magn Res Med 14: 68–78

Segebarth C, Belle V, Delon C, Massarelli R, Decety J, Le Bas JF, Decorps M, Benabid AL (1994) Functional MRI of the human brain: Predominance of signals from extracerebral veins. Neuro Report 5: 813–816

Thulborn KR, Waterton JC, Matthews PM, Radda GK (1982) Oxygenation dependence of the transverse relaxation time of water protons in whole blood at high field. Biochim Biophys Acta 714: 265–270

Author's address: Dr. J. Laubenberger, Radiologische Universitätsklinik, Abteilung Röntgendiagnostik, Hugstetter Strasse 55, D-79106 Freiburg, Federal Republic of Germany.

Measurement of the dopaminergic degeneration in Parkinson's disease with [123I]β-CIT and SPECT

Correlation with clinical findings and comparison with multiple system atrophy and progressive supranuclear palsy

T. Brücke[1,2], **S. Asenbaum**[1,2], **W. Pirker**[1], **S. Djamshidian**[1], **S. Wenger**[1], **Ch. Wöber**[1], **Ch. Müller**[1], and **I. Podreka**[3]

University Clinic for [1]Neurology and [2]Nuclear Medicine, Vienna, [3]Rudolfstiftung Hospital, Vienna, Austria

Summary. The cocaine derivative [123I]β-CIT binds with high affinity to dopamine uptake sites in the striatum and can be used to visualize dopaminergic nerve terminals in vivo in the human brain with SPECT. It has been validated that the calculation of a simple ratio of specific/nondisplaceable binding during a period of binding-equilibrium in the striatum about 20 hrs after bolus injection of the tracer gives a strong and reliable index of the binding potential of dopamine uptake sites. Previous studies have shown that the dopaminergic deficit in patients with Parkinson's disease (PD) can clearly be visualized and quantified using this method. Our own results in a group of 113 patients with PD demonstrate a 45% loss of striatal [123I]β-CIT binding in comparison to age corrected control values. Highly significant correlations of SPECT findings with clinical data obtained from the UPDRS rating scale such as akinesia, rigidity, axial symptoms and activities of daily living are demonstrated, while no correlation is found with tremor. The signal loss in a region comprising the whole striatum ranges from 35% in Hoehn/Yahr stage I to over 72% in stage V and is highly significantly correlated to the different stages of disease severity. A comparison of [123I]β-CIT binding in the striatum contralaterally and ipsilaterally to the affected body side in 29 patients with hemiparkinson shows a loss of striatal binding of 41% contralaterally and 30% ipsilaterally. Results from subregional analyses in caudate and putamen show relative sparing of the caudate nucleus in PD. Data in 9 patients with multiple system atrophy (MSA) and 4 patients with progressive supranuclear palsy (PSP) are similar to the findings in PD although the differences between caudate and putamen are somewhat less marked.

These data demonstrate that the dopaminergic nerve cell loss in PD and other disorders with a dopaminergic lesion can be quantified with [123I]β-CIT and SPECT and that hopefully a preclinical or very early diagnosis is made possible. Such studies might also open the way for a better evaluation of neuroprotective strategies in PD. It does not seem to be possible however to differentiate PD and MSA or PSP with this method in individual cases.

Introduction

Parkinson's disease (PD) is a slowly progressing neurodegenerative disorder
with a loss of dopaminergic neurons in the substantia nigra which leads to a
loss of dopaminergic nerve endings and to a marked reduction of the
dopamine content in the striatum. Dopamine transporters are located presyn-
aptically on the plasma membrane of dopaminergic terminals and are thus lost
in the process of degeneration (Pimoule et al., 1983; Janowsy et al., 1987;
Maloteaux et al., 1988). Idiopathic PD is characterized by symptoms such as
resting tremor, akinesia, rigidity and postural instability. Typically symptoms
start asymmetrically on one body-side, gradually affect both sides and usually
respond well to levodopa. Although these symptoms are very characteristic
and sometimes make it possible to diagnose this disorder by just watching a
patient as it is described in James Parkinson's original monography (1817) the
differential diagnosis with other extrapyramidal syndromes often can be very
difficult. In a clinico-pathological study it was recently reported that the
clinical diagnosis of PD was only correct in about 80% of the cases even when
strict diagnostic criteria had been used (Hughes et al., 1992). It can only be
speculated that this number would probably be lower in patients who are
diagnosed by physicians who are less familiar with the disorder. Therefore
there is a need for an objective diagnostic test for PD. Because the evaluation
of disease progression based on clinical neurological examinations is often
difficult due to the effects of antiparkinsonian treatment there is also a need
for an objective measure of the dopaminergic nerve cell loss in vivo to
evaluate possible neuroprotective strategies.

   One way to visualize and measure dopaminergic nerve endings in the brain
in vivo is to label the dopamine transporter either with positron- or with single
photon emitting ligands. In vitro studies in postmortem human brain samples
of PD patients had shown a loss of dopamine transporter binding with
[^{3}H]cocaine and [^{3}H]GBR-12935 (Pimoule et al., 1983; Janowsky et al., 1987).
These findings could be replicated in vivo with the PET ligand
[^{11}C]nomifensine (Aquilonius et al., 1987; Tedroff et al., 1988). Recently a
group of cocaine analogs with very high affinity for the dopamine transporter
and less nonspecific binding was described (Madras et al., 1989). PET studies
with the [^{11}C] labeled tropane analog WIN 35428 (CFT) also demonstrated
reduced binding in patients with PD (Frost et al., 1993). 2-β-carbomethoxy-3-
β-(4-iodophenyl)-tropane (β-CIT, RTI 55) is an iodinated analog of the
originally described fluoro-derivative CFT with further increased affinity for
monoamine transporters (Boja et al., 1991; Neumeyer et al., 1991) which has
been extensively characterized in animal experiments (Innis et al., 1991;
Scheffel et al., 1992; Shaya et al., 1992; Laruelle et al., 1993; Laruelle et al.,
1994a) and its binding studied in postmortem human brain samples (Farde et
al., 1994; Staley et al., 1994). SPECT studies with [^{123}I] labelled β-CIT in
patients with PD have shown that it is possible to visualize and quantify the
loss of dopaminergic nerve endings in this disorder and that the results
correlate well with clinical measures of disease severity, motor impairment
and asymmetry of symptomatology (Kuikka et al., 1993; Brücke et al., 1993,

1994, 1995; Innis et al., 1993; Marek et al., 1994, 1996; Seibyl et al., 1995; Rinne et al., 1995; Asenbaum et al., 1997a). The aim of the present study was to extend our findings in a larger number of patients, to correlate subscores of different symptoms with [^{123}I]β-CIT binding data and to test its potency in the differential diagnosis of parkinsonian syndromes.

Subjects and methods

Patients and controls

The control group consisted of 13 healthy volunteers and patients with peripheral neurological disorders [7 volunteers, 6 patients; 7 male, 6 female; age: 51.1 ± 20.4 (26–75) yrs].

Patients: A large group of patients with PD was studied [n = 113; 75 male, 38 female; age: 65.2 ± 11.5 (39–85) yrs]. Patients fulfilled the clinical criteria for the diagnosis of PD (Hughes et al., 1992) and were responsive to levodopa. In 80 patients enough clinical information was available to determine disease severity according to Hoehn and Yahr (Hoehn and Yahr, 1967). 29 patients were in stage I (mean age 62.5 yrs), 8 in stage II (mean age 60.3 yrs), 27 in stage III (mean age 68.2 yrs), 14 in stage IV (mean age 67.9 yrs) and 2 in stage V (mean age 74 yrs). Of these patients 61 were rated with the Unified Parkinson's Disease Rating Scale (UPDRS) (Fahn et al., 1987) at the time of the SPECT examination. Patients were allowed to take their antiparkinsonian medication on the day of tracer administration with the exception of L-deprenyl and benztropine which were stopped at least 18 hours before. Clinical ratings were performed in the early afternoon in patients on antiparkinsonian medication.

From the group of PD patients 14 with mild to moderate disease severity [PD 1: H/Y 2.1 ± 1.0 (1–3); mean age: 60.4 yrs] and 8 with moderate to severe disease [PD 2: H/Y 3.1 ± 0.6 (2–4); mean age: 64.6 yrs] were analyzed subregionally in the caudate nucleus and the putamen to compare them with a group of 13 patients with multiple system atrophy (MSA; n = 9) and progressive supranuclear palsy (PSP; n = 4) (H/Y 3–4; mean age: 61.8 yrs) and with 11 of the controls (mean age: 49.2 yrs) in whom caudate and putamen were also analyzed seperately. The 9 patients with MSA consisted of: 6 striato-nigral degeneration, 2 Shy-Drager and 1 olivo-ponto-cerebellar atrophy type. MSA patients fulfilled clinical criteria of probable MSA (Quinn et al., 1994) and all but one also had SPECT studies with [^{123}I]-IBZM or [^{123}I]-epidepride to examine the striatal D2 receptor status (W. Pirker in preparation). All PSP patients had vertical down- and upgaze palsy.

Demographic data of the control and patient groups are given in Table 1.

The study was approved by the local ethics committee and informed consent was obtained from each subject.

Table 1. Clinical data of controls and patients

Groups	n	age (range)	m/f
Controls	13	51.1 ± 20.4 (26–75)	7/6
Parkinson (stage 1–5)	113	65.2 ± 11.5 (39–85)	75/38
MSA	9	61.8 ± 11.1 (44–82)	6/7
PSP	4		

SPECT study

After blockade of thyroid uptake with 600 mg sodium perchlorate orally 30 minutes before tracer application subjects received a mean dose of 140 MBq (3.8 mCi) (range 104–222 MBq; 2.6–5.4 mCi) of [^{123}I]β-CIT i.v. as a bolus.

SPECT-studies were performed using a three-headed rotating scintillation camera (Siemens Multispect 3, FWHM 9 mm) equipped with medium-energy collimators. The subject's head was positioned in the head holder by means of a crossed laser beam system. The patient and all controls were studied 20 hrs after i.v. injection of the tracer because earlier studies from our group had shown a binding equilibrium at this time-point (Brücke et al., 1994; Asenbaum et al., 1996a). Imaging lasted 40 minutes. For each scan a total of 180 frames (40 sec per frame) was obtained in a step and shoot mode. 3.5 mm thick cross sections oriented parallel to the cantho-meatal plane were reconstructed by filtered back projection (Butterworth filter cutoff frequency 0.7, order 7) in 128×128 matrices. Attenuation correction was performed with a uniform attenuation coefficient of 0.12/cm after manually drawing an ellipse around the head contour. Irregular regions of interest (ROIs) were drawn manually on single slice views in areas corresponding to the left and right striatum (size: 40 to 45 pixels each) and the cerebellar hemispheres on either side (55 to 60 pixels each). All ROIs were drawn with the help of a brain atlas by one and the same examiner. Counts in striatal regions were calculated in several consecutive 3.5 mm thick axial slices and the highest value for each striatum was taken to avoid tilting errors. The two striatal ROIs were pooled together and the average counts/ pixel were calculated. Cerebellar ROIs were drawn on the slice of best visualization, usually 10 slices below the maximal activity in the striatum and in the two adjacent slices. Left and right cerebellar values were pooled and the average values from the three slices taken as reference region. Cerebellar activity was assumed to represent nonspecific bound and free radioactivity because it is known that the cerebellum has a very low density of dopamine and 5-HT transporters. For analysis a ratio was calculated between average count-rates in the striatum and the cerebellum. This ratio minus 1 represents specific/nondisplaceable binding and during a period of binding equilibrium it is directly related to the binding potential. The data of the patients were compared with the age corrected control values obtained from the regression analysis of the data from the control group according to the formula: $y = -0.036 \times +9.067$ (y = age corrected ratio; x = age of the patient). The percent deviation of each patients ratio from this age corrected control value was calculated.

Subregional analysis

In the multiple system atrophy (MSA) and the progressive supranuclear palsy (PSP) patients and in the PD patients (PD 1 and PD 2) and 11 controls which were used for comparison irregular ROIs in the caudate nucleus and the putamen (20 pxls each) were drawn in addition to ROIs which comprised the whole striatum. Caudate/putamen ratios were calculated: total countrate in the caudate nucleus divided by total countrate in the putamen.

An asymmetry index between left and right striatal [^{123}I]β-CIT binding was calculated in the same patients in whom a subregional analysis was performed: Average striatal (Left + right / 2) specific/nondisplaceable ratio minus ratio in the striatum with lower binding divided by average striatal specific/nonspecific ratio times 100.

Values of MSA and PSP patients were pooled together for statistical analysis.

Statistics

Binding ratios of controls and patients were compared with 2-tailed Student's t-test for unpaired samples, contra- and ipsilateral striatum in hemiparkinsonian patients with t-test for paired samples. Clinical rating scores and disease severity scores were correlated to the [^{123}I]β-CIT binding ratios with regression analysis. One way ANOVA was used to evaluate differences between the PD groups with different disease severity and between contra- and ipsilateral striatum in patients with hemiparkinson and controls and also for data from the subregional analysis.

Values are given as mean ± SD (range). p < 0.05 was considered statistically significant.

Results

Figure 1 gives the results in the total group of PD patients (n = 113). There was a highly significant difference of the specific/nondisplaceable binding ratio in the whole striatum (left and right pooled together) in the PD group compared to the controls (n = 13) with almost no overlap of the data (3.71 ± 1.15 vs 7.22 ± 1.08; p < 0.0001; PD vs controls). The reduction in percent of the age corrected value was 45%.

A significant age-dependent decline of [^{123}I]β-CIT binding was found in the control goup (4.9% of the mean ratio per decade; R = –0.76; p = 0.01) and was also seen less significantly in the PD group (R = 0.2; p = 0.036).

In hemiparkinsonian patients (n = 29) the comparison of specific/nondisplaceable binding in the whole striatum contralaterally and ipsilaterally to the affected body side revealed a highly significant difference (4.05 ± 0.76 and 4.79 ± 0.096; p = 0.0001; contra- and ipsilaterally respectively). Ratios were

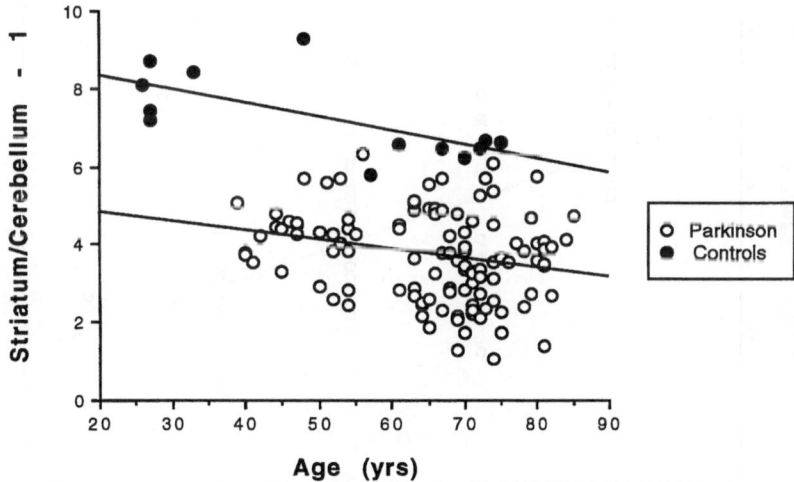

Fig. 1. Specific/nondisplaceable (Striatum/Cerebellum – 1) binding ratio of [^{123}I]β-CIT in the whole striatum (left and right pooled together) in the control group (filled circles, n = 13) and in patients with Parkinson's disease (open circles, n = 113) plotted against age. A marked reduction of the ratio is seen in the patient group (3.71 ± 1.15 vs. 7.22 ± 1.08; p < 0.0001; patients vs controls) with overlap only in a few cases with hemiparkinson. A significant age-dependent decline of [^{123}I]β-CIT binding was found in the control group (4.9% of the mean ratio per decade; R = –0.76; p = 0.01) and was also seen less significantly in the PD group (R = 0.2; p = 0.036)

also significantly reduced on both sides in comparison to controls (p = 0.0001). The percent reduction in comparison to age corrected values was 41% contralaterally and 30% ipsilaterally (Fig. 2).

Figure 3 shows the correlation of [^{123}I]β-CIT binding ratios and disease severity according to Hoehn and Yahr. A highly significant negative correlation was

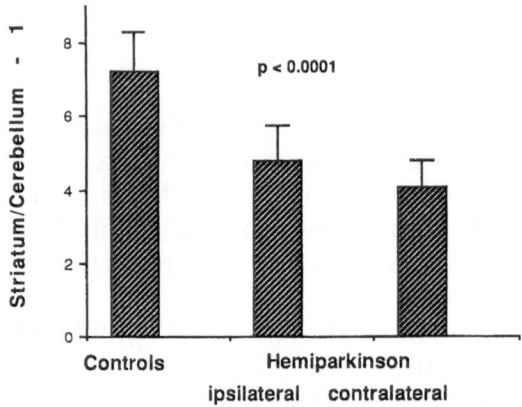

Fig. 2. Specific/nondisplaceable striatal binding ratio of [^{123}I]β-CIT in controls and patients with hemiparkinson (n = 29) ipsi- and contralaterally to the affected body side. Reduction on the ipsilateral side 30% and on the contralateral side 41% of age corrected control values (4.79 ± 0.096, 4.05 ± 0.76; p = 0.0001; mean ± SD)

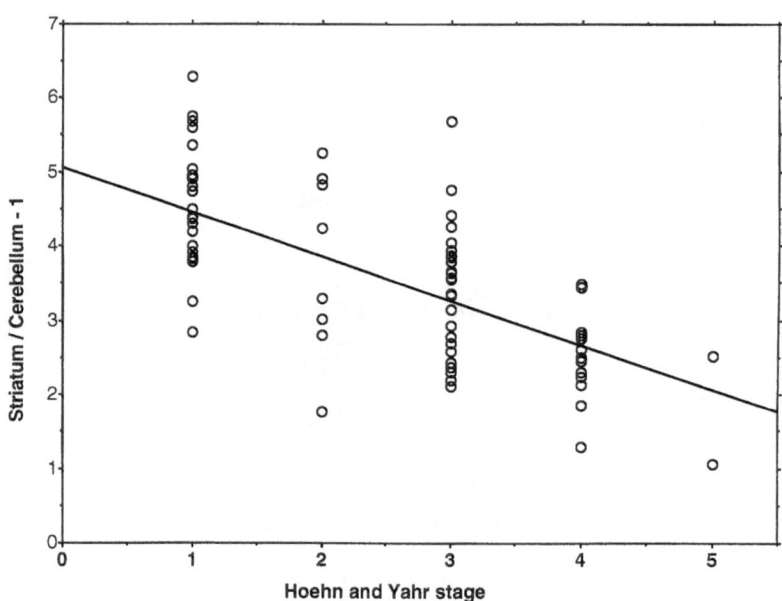

Fig. 3. Correlation of specific/nondisplaceable striatal binding of [^{123}I]β-CIT and disease severity according to Hoehn and Yahr (n = 80). A highly significant negative correlation is found (R = –0.66; p = 0.0001). ANOVA also reveales a highly significant difference between groups (F = 15.9; p = 0.0001; group 1 vs 3, 4 and 5, group 2 vs 4 and 5, group 3 vs 4 and 5). Reductions to age corrected values are 35% in stage 1, 46% in stage 2, 48% in stage 3, 62% in stage 4 and 72% in stage 5

found (R = –0.66; p = 0.0001). ANOVA also revealed a highly significant difference between groups (F = 15.9; p = 0.0001; group 1 vs 3, 4 and 5, group 2 vs 4 and 5, group 3 vs 4 and 5). Reductions to age corrected values were 35% in stage 1, 46% in stage 2, 48% in stage 3, 62% in stage 4 and 72% in stage 5.

Comparing clinical findings assessed with the UPDRS rating scale with [123I]β-CIT binding ratios revealed highly significant negative correlations with rigidity (R = –0.38; p = 0.0027), akinesia (R = –0.38; p = 0.0025) and axial symptoms (R = –0.52; p = 0.0001) but no correlation with tremor scores (R –0.15; n.s.) (Fig. 4). Total UPDRS motor scores and activities of daily living scores were also highly negatively correlated with [123I]β-CIT binding (R = –0.42; p = 0.0007 and R –0.55; p = 0.0004 respectively) (Fig. 5).

Striatal [123I]-β-CIT binding is markedly reduced in MSA and PSP patients. The overall reduction is about the same (–59% in our patient group) as in patients with PD with a similar degree of disability and somewhat more pronounced than in patients with mild disease (PD 2: –54%; PD 1: –48%).

The average reduction in the caudate nucleus was –53% in MSA and PSP

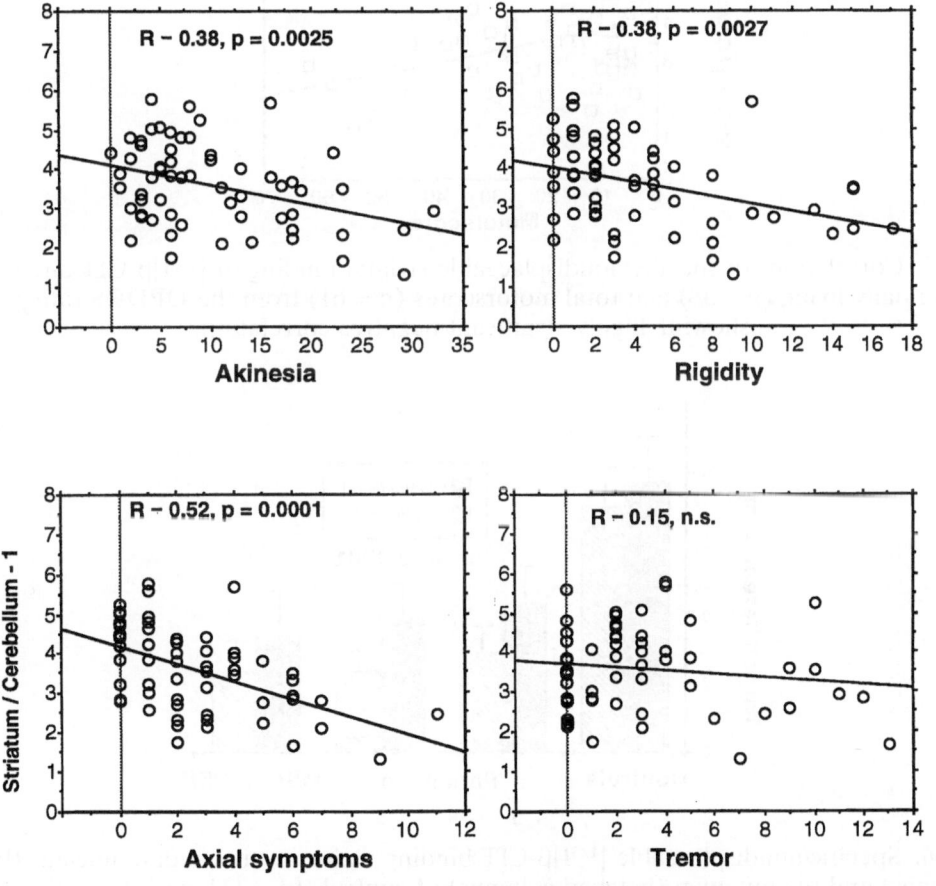

Fig. 4. Correlations of specific/nondisplaceable striatal binding of [123I]β-CIT and different subscores of the UPDRS rating scale for Parkinson's disease (n = 61). Highly significant negative correlations with akinesia, rigidity and axial symptoms but no correlation with tremor scores are found

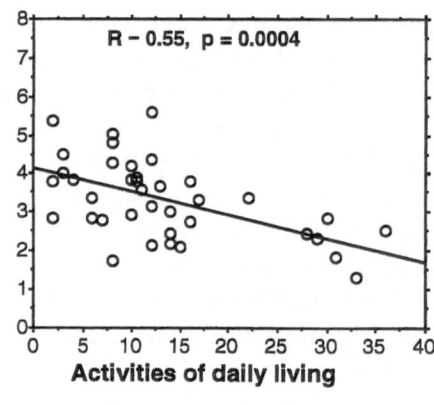

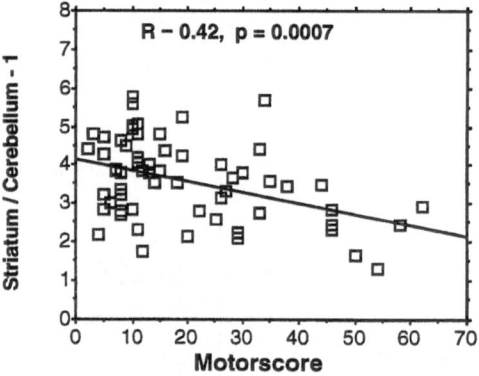

Fig. 5. Correlations of specific/nondisplaceable striatal binding of [^{123}I]β-CIT and activities of daily living (n = 38) and total motorscores (n = 61) from the UPDRS rating scale showing highly significant negative correlations

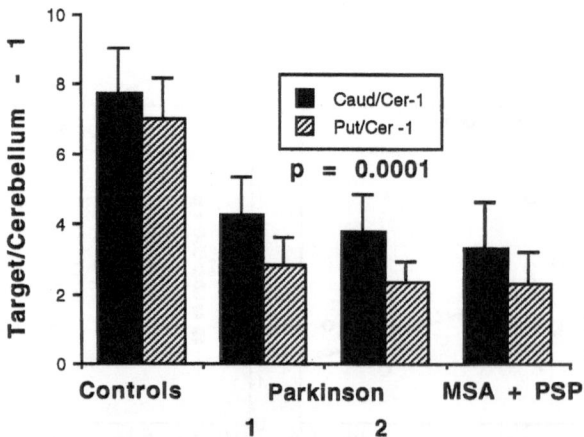

Fig. 6. Specific/nondisplaceable [^{123}I]β-CIT binding ratios in the caudate nucleus (black columns) and the putamen (hatched columns) of controls (n = 11), patients with mild to moderate (PD 1, n = 14) and moderate to severe Parkinson's disease (PD 2, n = 8) and patients with MSA or PSP (n = 13). A highly significant reduction in putamen and caudate nucleus is seen in all groups the reduction being more marked in the putamen (p = 0.0001). No significant difference is found between PD1, PD2 and MSA/PSP groups. Mean ± SD

patients and –40% and –45% in the PD 1 and PD 2 group respectively. Reduction in the putamen was –64% in MSA and PSP patients and –54% and –63% in the PD 1 and PD 2 group respectively (numbers calculated for age corrected control values) (Fig. 6).

There is a high degree of asymmetry of striatal [^{123}I]β-CIT binding between left and right hemisphere in PD patients and MSA and PSP patients in comparison to controls. Although there seems to be a trend to lower asymmetry in MSA and PSP patients this does not reach the level of significance in our patient population (Fig. 7).

The caudate/putamen ratio of [^{123}I]β-CIT binding is increased in PD patients as well as in MSA and PSP. This increase is somewhat less marked in the latter group but again this is not statistically significant (Fig. 8).

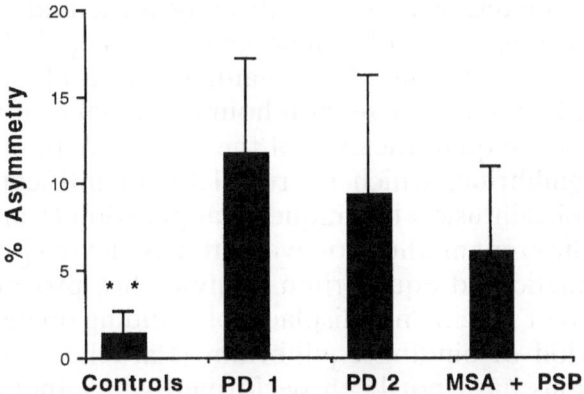

Fig. 7. Asymmetry index of [^{123}I]β-CIT binding between left and right striatum (for details see methods) in 11 controls, PD1, PD2 and MSA/PSP patients. Although there seems to be a trend to lower asymmetry in MSA and PSP patients compared with PD patients this does not reach the level of significance. Compared with controls the asymmetry index is significantly higher in all groups (p = 0.003). Mean ± SD

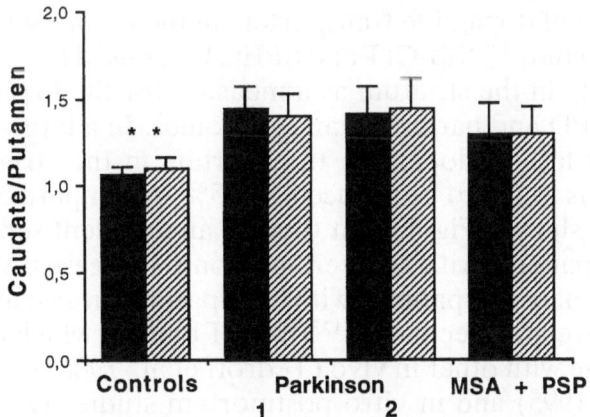

Fig. 8. Caudate/putamen ratios of [^{123}I]β-CIT binding on the right (black columns) and on the left side (hatched columns) are increased in PD1 and PD2 patients as well as in MSA and PSP (p = 0.0001) in comparison with controls. This increase is somewhat less marked in the latter group but this is not statistically significant. Mean ± SD

Discussion

β-CIT is a ligand with high affinity to dopamine- and serotonin uptake sites (Boja et al., 1991, 1992; Neumeyer et al., 1991; Laruelle et al., 1994a). In vivo studies in primates have shown that its binding in the striatum is practically exclusive to dopamine transporters (Laruelle et al., 1993). Studies from our own group demonstrated no reduction of striatal [^{123}I]β-CIT binding in patients treated with the selective serotonin reuptake inhibitor citalopram thus indicating that the striatal signal after i.v. application of this ligand is due to the dopamine uptake sites and is not influenced by serotonin transporters to a measurable degree (Pirker et al., 1995). Kinetic studies have shown that [^{123}I]β-CIT reaches its maximal specific binding in the striatum around 20 hours after injection (Brücke et al., 1993; Innis et al., 1993). This slow kinetic has been considered a disadvantage for its clinical use and has prompted the search for derivatives with faster brain uptake (Seibyl et al., 1996b; Malison et al., 1996). However, β-CIT has the advantage of reaching a true binding equilibrium which is stable over several hours and makes it possible to use a simple ratio method for quantification of the results. With most other ligands such a binding equilibrium which is crucial for quantification can only be achieved with a bolus/infusion technique of application (Carson et al., 1993). A comparison of different methods of evaluation of in vivo [^{123}I]β-CIT binding with graphical, kinetic and equilibrium analyses has proven the validity of using a simple ratio of specific/nondisplaceable binding on day 2 after application during a period of binding equilibrium (Laruelle et al., 1994b). Such thorough evaluations have not been performed with other CIT derivatives. However, it has been suggested that the loss of dopamine transporters in PD might be overestimated with ligands with faster uptake and faster washout which do not reach a binding equilibrium such as FP-CIT the fluoro-propyl derivative of CIT (Seibyl et al., 1996b). Furthermore a good reproducibility of [^{123}I]β-CIT SPECT measurements in test-retest experiments was demonstrated which suggests that this is a reliable and reproducible method for the in vivo measurement of dopamine transporters in the brain (Seibyl et al., 1996a).

In the present study [^{123}I]β-CIT and SPECT was used to examine dopamine transporter density in the striatum as a measure for the loss of dopaminergic nerve endings in PD and parkinsonian syndromes. In a large group of PD patients the overall loss of dopamine transporters in the striatum (both sides pooled together) is found to be reduced by 45% in comparison to the age corrected value. It is shown (Fig. 1) that there is an excellent separation between controls and PD patients with an overlap in only two cases. Visual and subregional analysis in these two patients with hemiparkinson also made a distinction possible. The age-related decline of [^{123}I]β-CIT binding which is seen in the control group is in line with other in vivo (Tedroff et al., 1988; Volkow et al., 1994; van Dyck et al., 1995) and in vitro postmortem studies (Zelnik et al., 1986; DeKeyser et al., 1990) and probably reflects an age related loss of dopaminergic neurons. A less significant age-effect is also seen in the PD group.

Several previous SPECT studies with [^{123}I]β-CIT have shown a reduction of striatal binding in patients with PD (Kuikka et al., 1993; Brücke et al., 1993;

Innis et al., 1993) and also a good correlation with clinical parameters (Brücke et al., 1994, 1995; Marek et al., 1994, 1996; Seibyl et al., 1995; Rinne et al., 1995; Asenbaum et al., 1996a). In the present study the severity of the disease and of several symptoms such as akinesia, rigidity, axial symptoms and tremor were compared with the SPECT findings. Excellent correlations were found with all symptoms except tremor which is in line with a study by Seibyl et al. (1995). This suggests that the reduction of striatal $[^{123}I]\beta$-CIT binding is a good measure for the loss of nigro-striatal dopaminergic neurons which is the cause for akinesia and rigidity but that tremor is not directly related to this dopaminergic lesion. In fact postmortem studies in patients with tremor dominant PD apparently show that dopaminergic neurons in the substantia nigra pars compacta are less affected than in cases with akinetic-rigid subtypes and that neurons in the retrorubal field (A8) are more severely degenerated in these patients (K. Jellinger; personal communication). Correlations with clinical parameters such as disease severity, asymmetry of symptoms and severity of bradykinesia and in vivo indices of dopaminergic function have also been described before in $[^{18}F]$-DOPA PET studies in smaller numbers of PD patients (Leenders et al., 1990; Brooks et al., 1990; Eidelberg et al., 1990; Otsuka et al., 1991). From the subregional analysis of $[^{123}I]\beta$-CIT binding in caudate and putamen in patients with mild to moderate and moderate to severe disease a putaminal loss of 54% and 63% of dopamine terminals respectively is calculated from the present study. Almost exactly the same numbers have been published for the loss of $[^{18}F]$-DOPA influx constants in the putamen in PD patients (Brooks et al., 1990). These numbers contrast with postmortem studies of dopamine concentrations in the striatum which have described a reduction of 90% and more (Bernheimer et al., 1973; Kish et al., 1988). An explanation for this discrepancy may be that many dopaminergic neurons may still be morphologically intact and may also have a relatively preserved activity of aromatic amino acid decarboxylase which is measured with $[^{18}F]$-DOPA, but a marked decrease of tyrosine hydroxylase the rate limiting enzyme of dopamine synthesis. In line with this hypothesis are postmortem results which prove that the percentage of surviving dopaminergic neurons in the substantia nigra of PD patients is higher than expected from measurements of dopamine concentration in the striatum (German et al., 1989; Goto et al., 1989; Rinne et al., 1989). This offers hope for possible neuroprotective treatments in PD which will have a better chance of success if a larger proportion of dopaminergic neurons is still intact in patients in whom the disease is already clinically manifest.

There are some possible confounding factors in the validity of $[^{123}I]\beta$-CIT SPECT as a measure of the dopaminergic nerve cell loss in substantia nigra. One is the possibility that regulatory changes of dopamine transporters might accompany changes in the concentration of dopamine in the synaptic cleft (Kilbourn et al., 1992; Wilson et al., 1994; Malison et al., 1994). It might be conceivable that a compensatory decrease of dopamine transporters accompanies a loss of dopamine in PD. This would lead to an overestimation of the nerve cell loss in in vivo studies of the dopamine transporter. Another possibility is that axonal sprouting of nigrostriatal dopaminergic axons might occur in the process of degeneration which might affect SPECT results in the opposite way. Different subpopulations of dopaminergic neurons in the midbrain also seem to express synaptic dopamine transporters in differing amounts and it has been hypothesized that neurons with higher expression might be more susceptible to the degenerative process in PD (Pifl, 1996). Thus neurons with a primary lower expression rate of dopamine transporters would be more likely to survive which would again lead to an overestimation of the nerve cell loss in [^{123}I]β-CIT studies. The high correlation of clinical symptoms and SPECT results in the present and previous studies (Brücke et al., 1994, 1995; Marek et al., 1994, 1996; Rinne et al., 1995; Seibyl et al., 1995; Asenbaum et al., 1996a) nevertheless suggest that [^{123}I]β-CIT is a good measure for the degree of dopaminergic nigro-striatal degeneration.

The second part of our study deals with the potency of [^{123}I]β-CIT SPECT in the differential diagnosis of PD. It has been shown by our group that patients with essential tremor can clearly be delineated from PD and that these patients have normal [^{123}I]β-CIT binding in the striatum (Asenbaum et al., 1996b). In some cases of essential tremor differential diagnosis with patients with tremor dominant PD can be difficult clinically. Thus [^{123}I]β-CIT SPECT can help to establish the right diagnosis in these cases and to save them from unnecessary drug treatment. It has also been shown that patients with focal dystonias have normal striatal dopamine transporter binding (Asenbaum et al., 1996b; Naumann et al., in preparation) and that patients with vascular encephalopathy with gait problems who sometimes resemble PD ("lower body PD") can also easily be distinguished from idiopathic PD (own unpublished observations). In the study presented here we were interested if [^{123}I]β-CIT SPECT could be helpful to seperate patients with MSA and PSP from patients with PD. In spite of clear pathological differences between these disorders and in spite of clinical hallmarks (Quinn et al., 1994) these akinetic rigid syndromes can often cause diagnostic difficulties especially in their early clinical stages (Brooks et al., 1990). In PD the dopaminergic cells in the ventrolateral tier of the substantia nigra pars compacta seem to be more susceptible to the degenerating process (Bernheimer et al., 1973; Goto et al., 1989). These cells project mainly to the putamen which explains why dopamine depletion in the putamen is more marked in PD (Bernheimer et al., 1973; Kish et al., 1988). In patients with MSA heterogenous findings have been reported with a more diffuse lesion of the substantia nigra in some cases and a cell loss resembling that in PD in others (Goto et al., 1989). A more global involvement of the substantia nigra pars compacta has also been described in PSP patients (Jellinger et al., 1980). This is in good agreement with postmortem measurements of dopamine concentrations in caudate nucleus and putamen with a similar degree of dopamine depletion in these two regions in cases with PSP (Kish et al., 1985; Ruberg et al., 1985). In vivo studies of dopaminergic function in PSP patients with PET showed equally severe impairment of [^{18}F]-DOPA uptake in putamen and caudate while a more heterogenous picture was found in MSA patients with values between PD and PSP. PD patients, as expected, showed a more severe involvement of the putamen (Brooks et al., 1990). In the present study a predominant affection of the putamen was also seen in almost all patients with PD in the visual evaluation of the images and was confirmed in the subregional analysis. Although [^{123}I]β-CIT binding in the caudate nucleus was somewhat more affected in the MSA/PSP group a significant difference from PD patients was not evident. Also there was no difference between MSA and PSP patients. A significantly larger difference between left and right striatum in comparison to the control group was not only found in PD patients but also in the MSA/PSP group so that differences in the degree of asymmetry did not help in the differential diagnosis either. Although it is possible that a different pattern in the affection of [^{123}I]β-CIT binding might be found in a larger sample of MSA/PSP patients, it does not seem to be possible to make a distinction in individual cases on the basis of this examination alone. In these patients a combination of clinical findings with pre- and postsynaptic dopaminergic imaging and MRI scans will probably be the most accurate way to establish the right diagnosis.

In conclusion our results show that [^{123}I]β-CIT and SPECT is a sensitive and reliable method for the in vivo measurement of dopaminergic function which can quantitate the degree of dopaminergic degeneration in PD and other parkinsonian syndromes. Findings are highly correlated with clinical symptoms and with measures of disease severity. [^{123}I]β-CIT and SPECT can be used as a diagnostic test in PD which can help in the differential diagnosis but it does not seem to be possible to clearly distinguish MSA and PSP in individual cases from PD with this method alone.

References

Aquilonius SM, Bergström K, Eckernäs SA, Hartvig P, Leenders KL, Lundquist H, Antoni G, Gee A, Rimland A, Uhlin J (1987) In vivo evaluation of striatal dopamine reuptake sites using 11C-nomifensine and positron emission tomography. Acta Neurol Scand 76: 283–287

Asenbaum S, Brücke T, Pirker W, Podreka I, Angelberger P, Wenger S, Wöber C, Müller C, Deecke L (1997a) Imaging of dopamine transporters with [123I]β-CIT and SPECT in Parkinson's disease. J Nucl Med (in press)

Asenbaum S, Brücke T, Pirker W, Wenger S, Podreka I, Deecke L (1997b) The application of [123I]β-CIT and SPECT in the diagnosis of movement disorders. Proceedings of the meeting on Neurospect, Antwerp

Bernheimer H, Birkmayer W, Hornykiewicz O, Jellinger K, Seitelberger F (1973) Brain dopamine and the syndromes of Parkinson and Huntington. J Neurol Sci 20: 415–455

Boja JW, Patel A, Carroll FI, Rahman MA, Philip A, Lewin AH, Kopajtic TA, Kuhar MJ (1991) [125I]RTI-55: a potent ligand for dopamine transporters. Eur J Pharmacol 194: 133–134

Boja J, Mitchell WM, Patel A, Kopajtic TA, Carroll FI, Lewin AH, Abraham P, Kuhar MJ (1992) High affinity binding of [125I]RTI-55 to dopamine and serotonin transporters in rat brain. Synapse 12: 27–36

Brooks DJ, Ibanez V, Sawle GV, Quinn N, Lees AJ, Mathias CJ, Bannister R, Marsden CD, Frackowiak RSJ (1990) Differing patterns of striatal 18F-Dopa uptake in Parkinson's disease, multiple system atrophy, and progressive supranuclear palsy. Ann Neurol 28: 547–555

Brücke T, Kornhuber J, Angelberger P, Asenbaum S, Frassine H, Podreka I (1993) SPECT imaging of dopamine and serotonin transporters with [123I]β-CIT.Binding kinetics in the human brain. J Neural Transm [GenSect] 94: 137–146

Brücke T, Asenbaum S, Pozzera A, Hornykiewicz S, Harasko-van der Meer C, Wenger S, Koch G, Pirker W, Wöber C, Müller C (1994) Dopaminergic nerve cell loss in Parkinson's disease quantified with [123I]ß-CIT and SPECT correlates with clinical findings. Mov Disord 9 (S1): 120

Brücke T, Asenbaum S, Pirker W, Pozzera A, Wenger S, Wöber C, Müller C, Angelberger P, Podreka I (1995) Quantification of the dopaminergic nerve cell loss in Parkinson's disease with [123I]β-CIT and SPECT. J Cereb Blood Flow Metab 15 [Suppl 1]: 37

Carson RE, Channing MA, Blasberg RG, Dunn BB, Cohen RM, Rice KC, Herscovitch P (1993) Comparison of bolus and infusion methods for receptor quantitation: application to [18F]Cyclofoxy and positron emission tomography. J Cereb Blood Flow Metab 13: 24–42

De Keyser JD, Ebinger G, Vauquelin G (1990) Age-related changes in the human nigrostriatal dopaminergic system. Ann Neurol 27: 157–161

Eidelberg D, Moeller JR, Dhawan V, Sidits JJ, Ginos JZ, Strother SC, Cedarbaum J, Greene P, Fahn S, Rottenberg DA (1990) The metabolic anatomy of Parkinson's disease: complementary [18F] fluorodesoxyglucose and [18F] fluorodopa positron emission tomographic studies. Mov Dis 5: 203–213

Fahn S, Elton R, Members of the Unified Parkinson's Disease Rating Scale development committee (1987) Unified Parkinson's disease rating scale. In: Fahn S, Marsden CD, Calne DB, Goldstein M (eds) Recent developments in Parkinson's disease. Macmillan Healthcare Information, Florham Park, New York, 2: 153–164

Farde L, Halldin C, Müller L, Suhara T, Karlsson P, Hall H (1994) PET study of [11C]beta-CIT binding to monoamine transporters in the monkey and human brain. Synapse 16: 93–103

Frost JJ, Rosier AJ, Reich SG (1993) Positron emission tomography imaging of the dopamine transporter with 11C-WIN 35428 reveals marked declines in mild Parkinson's disease. Ann Neurol 34: 423–431

German DC, Manaye K, Smith WK, Woodward DJ, Saper CB (1989) Midbrain dopaminergic cell loss in Parkinson's disease: computer visualization. Ann Neurol 26: 507–514

Goto S, Hirano A, Matsumoto S (1989) Subdivisional involvement of nigrostriatal loop in idiopathic Parkinson's disease and striatonigral degeneration. Ann Neurol 26: 766–770

Hoehn MM, Yahr MD (1967) Parkinsonism: onset, progression and mortality. Neurol 6: 253–259

Hughes AJ, Daniel SE, Kilford L, Lees AJ (1992) Accuracy of clinical diagnosis of idiopathic Parkinson's disease: a clinico-pathological study of 100 cases. J Neurol Neurosurg Psychiatry 55 (3): 181–184

Innis R, Baldwin R, Sybirska E, Zea Y, Laruelle M, Al-Tikriti M, Charney D, Zoghbi S, Smith E, Wisniewski G (1991) Single photon emission computed tomography imaging of monoamine reuptake sites in primate brain with [123I]CIT. Eur J Pharmacol 299: 369–370

Innis RB, Seibyl JP, Scanley BE, Laruelle M, Abi-Dargham A, Wallace E, Baldwin R, Zea-Ponce Y, Zoghbi S, Wang S (1993) Single photon emission computed tomographic imaging demonstrates loss of striatal dopamine transporters in Parkinson disease. Proc Natl Acad Sci USA 90: 11965–11969

Janowsy A, Vocci F, Berger P, Angel I, Zelnik N, Kleinman J, Skolnick P (1987) [3H]GBR-12935 binding to the dopamine transporter is decreased in the caudate nucleus in Parkinson's disease. J Neurochem 49: 617–621

Jellinger K, Riederer P, Tomanaga M (1980) Progressive supranuclear palsy: clinico-pathological and biochemical studies. J Neural Transm [Suppl] 16: 111–128

Kaufmann MJ, Madras BK (1991) Severe depletion of cocaine recognition sites associated with the dopamine transporter in Parkinson's diseased striatum. Synapse 49: 43–49

Kilbourn MR, Sherman PS, Pisani T (1992) Repeated reserpine administration reduces in vivo [18F]GBR 13119 binding to the dopamine uptake site. Eur J Pharmacol 216: 109–112

Kish SJ, Chang LJ, Mirchandani L (1985) Progressive supranuclear palsy: relationship between extrapyramidal disturbances, dementia, and brain neurotransmitter markers. Ann Neurol 18: 530–536

Kish SJ, Shannar K, Hornykiewicz O (1988) Uneven pattern of dopamine loss in the striatum of patients with idiopathic Parkinson's disease. Pathophysiological and clinical implications. N Engl J Med 318: 876–880

Kuikka JT, Bergström KA, Vanninen E, Laulumaa V, Hartikainen P, Länsimies E (1993) Initial experience with single photon emission tomography using iodine-123-labelled 2β-carbomethoxy-3β-(4-iodophenyl)tropane in human brain. Eur J Nucl Med 20: 783–786

Laruelle M, Baldwin RM, Malison RT, Zea-Ponce Y, Zoghbi SS, Al-Tikriti MS, Sybirska EH, Zimmermann RC, Wisniewski G, Neumeyer JL (1993) SPECT Imaging of dopamine and serotonin transporters with [123I]β-CIT: pharmacological characterization of brain uptake in nonhuman primates. Synapse 13: 295–309

Laruelle M, Giddings S, Zea-Ponce Y, Charney DS, Neumeyer JL, Baldwin RM, Innis RB (1994a) Methyl 3β-(4-[125I]Iodophenyl) Tropane-2β-Carboxylate in vitro binding to dopamine and serotonin transporters under "physiological" conditions. J Neurochem 62: 978–986

Laruelle M, Wallace E, Seibyl JP, Baldwin RM, Zea-Ponce Y, Zoghbi SS, Neumeyer JL, Charney DS, Hoffer PB, Innis RB (1994b) Graphical, kinetic and equilibrium analyses of in vivo [123I]β-CIT binding to dopamine transporters in healthy human subjects. J Cereb Blood Flow Metab 14: 982–994

Leenders KL, Salmon EP, Tyrrell P, Perani D, Brooks DJ, Sager H, Jones T, Marsden D, Frackowiak RSJ (1990) The nigrostriatal dopaminergic system assessed in vivo by positron emission tomography in healthy volunteer subjects and patients with Parkinson's disease. Arch Neurol 47: 1290–1298

Madras BK, Spealman RD, Fahey MA, Neumeyer JL, Saha JK, Milius RA (1989) Cocaine receptors labeled by 2β-carbomethoxy-3β-(4-fluorophenyl)tropane. Mol Pharmacol 36: 518–524

Malison RT, Wallace EA, Best S, Gandelman M, Zoghbi SS, Zea-Ponce Y, Baldwin RM, Charney DS, Hoffer PB, Kosten TR (1994) SPECT imaging of dopamine transporters in cocaine dependent and healthy control subjects with 123I-β-CIT. Soc Neurosci Abstr 20 (2): 1625

Malison RT, Vessotskie J, Kung MP, McElgin W, Romaniello G, Kim HJ, Goodman MM, Kung HF (1996) SPECT imaging of striatal dopamine transporters in non-human primates with [123I]IPT. J Nucl Med (in press)

Marek KL, Seibyl JP, Sandridge B, Fussell B, Smith EO, Baldwin RM, Zoghbi S, Hoffer PB, Innis RB (1994) SPECT imaging demonstrates striatal dopamine transporter loss in hemiparkinsonism. Soc Neurosci Abstr 20: 1780

Marek KL, Seibyl JP, Zoghbi S, Zea-Ponce Y, Baldwin RM, Fussell B, Carney DS, van Dyck C, Hoffer PB, Innis RB (1996) [123I]β-CIT-SPECT imaging demonstrates bilateral loss of dopamine transporters in hemi-Parkinson's disease. Neurology 46: 231–237

Neumeyer JL, Wang S, Milius RA, Baldwin RM, Zea-Ponce Y, Hoffer PB, Sybirska E, Al-Tikriti MS, Laruelle M, Innis RB (1991) [123I]2β-carbomethoxy-3β-(4-iodophenyl)tropane (β-CIT): high affinity SPECT radiotracer of monoamine re-uptake sites in brain. J Med Chem 34: 3144–3146

Niznik HB, Fogel EF, Fassos FF, Seeman P (1991) The dopamine transporter is absent in parkinsonian putamen and reduced in the caudate nucleus. J Neurochem 56: 192–198

Otsuka M, Ichiya Y, Hosokawa S, Kuwabara Y, Tahara T, Fukumura T, Kato M, Masuda K, Goto I (1991) Striatal blood flow, glucose metabolism and 18F-Dopa uptake: difference in Parkinson's disease and atypical parkinsonism. J Neurol Neurosurg Psychiatry 54: 898–904

Parkinson J (1817) An essay on the shaking palsy. Sherwood, Neely and Jones, London

Pifl C, Uhl GR (1996) Dopamine Transporter: State of the art in parkinsonism. Mov Dis 11 (S1): 22

Pimoule C, Schoemaker H, Javoy-Agid F, Scatton B, Agid Y, Langer JZ (1983) Decrease in [3H]cocaine binding to the dopamine transporter in Parkinson's disease. Eur J Pharmacol 95: 145–146

Pirker W, Asenbaum S, Kasper S, Walter H, Angelberger P, Koch G, Pozzera A, Deecke L, Podreka I, Brücke T (1995) β-CIT SPECT demonstrates blockade of 5HT-uptake sites by citalopram in the human brain in vivo. J Neural Transm [Gen Sect] 100: 247–256

Quinn N (1994) Multiple system atrophy. In: Marsden CD, Fahn S (eds) Movement disorders 3. Butterworths-Heinemann, London, pp 262–281

Rinne JO, Rummukainen J, Paljärvi L, Rinne UK (1989) Dementia in Parkinson's disease is related to neuronal loss in the medial substantia nigra. Ann Neurol 26: 47–50

Rinne JO, Kuikka JT, Bergström KA, Rinne UK (1995) Striatal dopamine transporter in different disability stages of Parkinson's disease studied with [123I] β-CIT SPECT. Parkinsonism and Related Disorders 1 (1): 47–51

Ruberg M, Javoy-Agid F, Hirsch E, Scatton B, Lheureux R, Hauw JJ, Duyckaerts C, Gray F, Morel-Maroger A, Rascol A (1985) Dopaminergic and cholinergic lesions in progressive supranuclear palsy. Ann Neurol 18: 523–529

Scheffel U, Dannals RF, Cline EJ, Ricaurte GA, Carrol FI, Abraham P, Lewin AH, Kuhar MJ (1992) [123/125I]RTI-55, an in vivo label for the serotonin transporter. Synapse 11: 134–139

Seibyl JP, Marek KL, Quinlan D, Sheff K, Zoghbi S, Zea-Ponce Y, Baldwin RM, Fussell B, Smith EO, Charney DS (1995) Decreased single photon emission computed tomographic [123I]β-CIT striatal uptake correlates with symptom severity in Parkinson's disease. Ann Neurol 38: 589–598

Seibyl JP, Laruelle M, van Dyck CH, Wallace E, Baldwin RM, Zoghbi S, Zea-Ponce Y, Neumeyer JL, Charney DS, Hoffer PB (1996a) Reproducibility of iodine-123-β-CIT SPECT brain measurement of dopamine transporters. J Nucl Med 37: 222–228

Seibyl J, Marek K, Sheff K, Neumeyer J, Innis R (1996b) Comparison of [123I]FP-CIT and [123I]β-CIT for SPECT imaging of dopamine transporters in Parkinson's disease. J Nucl Med 37: 133P

Shaya EK, Scheffel U, Dannals RF, Ricaurte GA, Carrol FI, Wagner HN, Kuhar MJ, Wong DF (1992) In vivo imaging of dopamine uptake sites in primate brain using single photon emission tomography (SPECT) and iodine-123 labeled RTI-55. Synapse 10: 169–172

Staley JK, Basile M, Flynn DD, Mash DC (1994) Visualizing dopamine and serotonin transporters in the human brain with the potent cocaine analogue [125]RTI-55: In vitro binding and autoradiographic characterization. J Neurochem 62: 549–556

Tedroff J, Aquilonius SM, Hartvig P, Lundqvist H, Gee AG, Uhlin J, Långström B (1988) Monoamine reuptake sites in the human brain evaluated in vivo by means of 11C-nomifensine and positron emission tomography: the effects of age and Parkinson's disease. Acta Neurol Scand 77: 192–201

van Dyck CH, Seibyl JP, Malison RT, Laruelle M, Wallace E, Zoghbi SS, Zea-Ponce Y, Baldwin RM, Charney DS, Hoffer PB (1995) Age-related decline in striatal transporter binding with iodine-123-β-CIT SPECT J Nucl Med 36: 1175–1181

Volkow ND, Fowler JS, Wang GJ (1994) Decreased dopamine transporters with age in healthy human subjects. Ann Neurol 36: 237–239

Wilson JM, Nobrega JN, Carroll ME, Niznik HB, Shannak K, Lac ST, Pristupa ZB, Dixon LM, Kish SJ (1994) Heterogeneous subregional binding patterns of 3H-WIN 35,428 and 3H-GBR 12,935 are differentially regulated by chronic cocaine self-administration. J Neurosci 14: 2966–2979

Zelnik N, Angel I, Paul SM, Kleinman JE (1986) Decreased density of human striatal dopamine uptake sites with age. Eur J Pharmacol 126: 175–176

Authors' address: Prof. Dr. T. Brücke, Universitätsklinik für Neurologie, Währinger Gürtel 18–20, A-1090 Wien, Austria.

rCBF SPECT in Parkinson's disease patients with mental dysfunction

S. Bissessur, G. Tissingh, E. Ch. Wolters, and **Ph. Scheltens**

Graduate School Neurosciences Amsterdam, Department of Neurology,
Academisch Ziekenhuis VU Amsterdam, The Netherlands

Summary.   Functional imaging of the brain using SPECT provides information correlative to the alterations of regional blood flow. In this paper we review the literature pertaining to SPECT in Parkinson's disease with and without dementia and depression. Parkinson's disease itself is not associated with a consistent pattern of cerebral blood flow alterations in the basal ganglia, but reduced parietal blood flow is more often reported. The heterogeneity of blood flow changes possibly reflects the multifactorial pathophysiology of the disease. In demented Parkinson's disease patients frontal hypoperfusion is often found or bilateral temporoparietal deficits, probably indicative of concommitant Alzheimer's disease. The SPECT studies undertaken in depressed patients with and without Parkinsons's disease show highly conflicting and inconsistent results, probably due to methodological and diagnostic flaws (especially the inclusion of demented Parkinson patients). Several lines of reasoning point to a prefrontal dysfunction and future SPECT studies are planned to study this region in non-demented Parkinson's disease patients with and without major depression.

Introduction

Functional imaging of the brain using single photon emission computed tomography (SPECT) provides information correlative to the alterations of regional cerebral blood flow (rCBF). Studying rCBF may be useful in normal subjects and patients with neurodegenerative diseases since in these groups a tight coupling between rCBF and metabolism has been shown (Frackowiak and Gibbs, 1983). In this paper we attempt to review the available information regarding rCBF SPECT in the diagnosis of Parkinson's disease (PD) with and without mental dysfunction, i.e. dementia and depression.

The principles of SPECT are detailed in the chapter of Tissingh et al., elsewhere in this volume. The-photon emission of these radiopharmaceuticals is in the optimal range for standard gamma cameras used for SPECT in most nuclear medicine departments. [123I] Amphetamine ([I123]IMP), Xenon 133 and [99mTc] hexamethylpropylenamine (HMPAO) have been widely used in static rCBF SPECT studies in recent years. Considerable differences in the mechanism of uptake between these radiotracers have emerged (De Jong and Van Royen, 1989). Because of the limited availability and higher costs of [I[123]]IMP and because of the shorter half-life and more favorable dosimetry of [^{99m}Tc]HMPAO, this has become the agent of choice for routine use of SPECT studying rCBF. The use of this radiopharmaceutical has also yielded a higher image resolution. Postprocessing analysis may provide semi-quantitative data in which the rCBF in the region of interest (ROI) is computed relative to the rCBF in any other part of the brain, usually the cerebellum or the occipital cortex, based on the assumption that the reference region is not affected in the disease under study.

In the following we will address the use of rCBF SPECT in demented and non-demented PD patients and psychiatric patients suffering from major depression. Based on the results in major depression the rationale for a study in depressed PD patients will be given.

Parkinson's disease

Dopamine depletion in the basal ganglia might result in consistent alterations of local functional activity, and consequently of rCBF. However, rCBF values in the basal ganglia of PD patients have been found to be reduced (Henriksen and Boas, 1985; Neumann et al., 1989), increased (Martin et al., 1984) or unchanged (Kuhl et al., 1984; Pizzolato et al., 1988) compared to control values. In addition, diffuse or focal cortical abnormalities have been reported (Lenzi et al., 1979; Kuhl et al., 1984; Globus et al., 1985; Tachibana et al., 1993). Pizzolato et al. (1988) studied 36 PD patients with varying degrees of severity of motor functioning and mental deterioration, and 10 controls. In general, they found no differences in HMPAO uptake in the basal ganglia between patients and controls, while left-right asymmetries were noted in some patients. Significant reductions of tracer uptake were found bilaterally in the parietal cortex. These deficits were more pronounced in patients with mental deterioration (as measured on the mini mental state examination), and in subjects who had been chronically treated with anticholinergic drugs. Costa et al. (1988) reported that in PD patients significant differences in caudate nucleus and thalamic rCBF were present depending on the intake of levodopa. Tachibana and co-workers (1993) investigated 19 non-demented PD patients and showed parietal hypoperfusion in some patients. Two conclusions may be drawn from these observations: (i) PD is not associated with a consistent pattern of rCBF alterations in the basal ganglia and (ii) a more consistent pattern of reduced parietal rCBF has been found, usually, but not always, associated with some evidence of cognitive decline. The heterogeneity of rCBF changes in PD possibly reflects the multifactorial pathophysiology of the disease.

Parkinson's disease with dementia

The incidence and prevalence of dementia in PD patients, as well as the exact cause, is still a matter of discussion. With respect to the etiology, the concept of overlap between PD and Alzheimer's disease (AD), two common illnesses in the elderly, seems an attractive hypothesis to explain many, but by no means all, cases of demented PD patients. On the other hand, some authors suggest that PD patients exhibit a subcortical dementia, similar to Huntington's disease or patients with Steel-Richardson-Olszewski syndrome, probably due to a disruption of the frontosubcortical connections. The alternative explanation of Lewy body accumulation in the cerebral cortex has also to be considered (Gibb, 1994).

SPECT studies investigating dementia in PD have limited themselves to study rCBF, probably because of the observed defects in AD (Weinstein et al., 1993). In the earlier mentioned study of Pizzolato et al. (1988) the observed parietal hypoperfusion areas resembled those described in AD. Similar findings were reported by Kuhl et al (1985), Spampinato et al. (1991) and Liu et al. (1992). The observed posterior hypoperfusion pattern in demented PD patients, however, was less extensive than in AD patients. In contrast, a study from our own institute (Kuiper et al., 1992) could not replicate these findings. The main explanantion may be the absence or presence of cortical symptoms in demented PD patients. In this respect, it is interesting that Sawada et al. (1992), using [123I]IMP, found frontal hypoperfusion in 12 of the 13 demented PD patients, while some degree of parietal hypoperfusion was present in 5 demented PD patients. Only parietal rCBF correlated significantly with the presence of cortical dysfunctions. This confirms the hypothesis that dementia in PD may be of subcortical origin, and that finding a hypoperfusion pattern similar to AD is probably caused by a selection bias, when patients with AD-like cortical dysfunctions are investigated. Wang et al. (1992) also showed that demented PD patients, according to the DSM-III-R criteria, had significantly lower HMPAO uptake in the frontal and basal ganglia regions than the non-demented PD patients. On the other hand, Jagust et al. (1992) showed very elegantly that in PD patients who performed most poorly on neuropsychological testing (but were not diagnosed as demented) showed lowest rCBF ratios in left and right temporal lobes. Frontal hypoperfusion was more linked to the presence of depression and poor performance on frontal lobe testing. More recently, Sperling et al. (1993) reported similar hypoperfusion patterns in 26 AD patients and 12 demented PD patients, suggesting the possibility of common cerebral pathology. However, details regarding the diagnosis and severity of dementia, as well as the presence or absence of cortical features were lacking, which severely limits the interpretation of this study.

Thus, SPECT imaging has provided evidence for frontal hypoperfusion in demented PD patients, with some indication for bilateral temporoparietal rCBF deficits, in a subgroup of demented PD patients, indicative of concomitant AD pathology. However, most SPECT studies, are methodologically flawed, both in the definition of dementia in PD (Kuiper et al., 1993), as well as lack of sufficient age matched controls, both mentally normal PD patients or healthy people. A selective bias has often been made by selecting PD patients with dementia, on the basis of a dementia syndrome with or without cortical features, thus in a way predicting the presence or absence of AD-like changes. In addition, the numbers of studied patients have often been small and, especially, SPECT studies lack quantative data. Computing ratios of rCBF, with the cerebellum as a reference, may be fairly safe in AD; for PD this has not been investigated thoroughly.

Parkinson's disease with depression

Cerebral localisation and hemispheric lateralisation of mental dysfunction have been the subject of several investigations. Studies of mood changes following brain lesions suggest an asymmetrical functional regulation of emotional behaviour: damage to the left hemisphere seems to be associated to depression, whereas lesions of the right hemisphere are associated with euphoria or indifference (Lishman 1968; Sackheim et al., 1982; Lipsey et al., 1983). rCBF SPECT studies have been tried to demonstrate cerebral dysfunction in depressed patients, but conflicting results have been found (Uytenhoef et al., 1983; Gur et al., 1984). In study by Delvenne et al. (1990) involving 38 patients with major depression, withdrawn from medication, and 16 controls Xenon-133 SPECT showed no significant differences, although the left sided cortical rim in the depressed patients was significantly lower in bipolar patients, compared to unipolar patients and controls. Mayberg et al. (1991) reported a systemic frontal-temporal pattern of hypoperfusion in patients with unipolar depression while Trzepacz et al. (1992) found frontal abnormalities and a global decreased blood flow in neurologic patients with major depression. In an attempt to replicate these findings Maes et al. (1993) undertook a [^{99m}Tc]HMPAO SPECT study in 43 unipolar depressed patients and 12 healthy controls. They found no differences at all between the groups in rCBF in different ROI's, nor between different depressive subtypes. However, PET studies (Baxter et al., 1985, 1989) showed consistent left anterolateral prefrontal cortex hypometabolism in major depression. There is also accumulating evidence for the role of the serotonin system in depression in PD (for review see Van Praag, 1994), which also links depression to the frontal lobes (Jolles, 1994). The latter was confirmed by Mayberg et al. (1990) who reported a selective hypometabolism in the inferior frontal lobe in depressed PD patients.

The above mentioned findings, and the tight relationship between depression and dementia in PD (Kuniyoshi et al., 1989; Tröster et al., 1995) warrant a study in which rCBF SPECT patterns of demented and non-demented and depressive PD patients, matched for disease duration, medication and age, are compared and correlated to the cognitive and depressive symptomatology. Such studies are currently ongoing in our institute.

References

Baxter LR Jr, Schwartz JM, Phelps ME, et al (1989) Reduction of prefrontal cortex glucose metabolism common to three types of depression. Arch Gen Psychiatry 46: 243–250
Baxter LR Jr, Phelps ME, Mazziotta JC, et al (1985) Cerebral metabolic rates for glucose in mood disorders: studies with positron emission tomography and fluorodeoxyglucose F18. Arch Gen Psychiatry 42: 441–447

Costa DC, Ell PJ, Burns A, Philpot M, Levy R (1988) CBF Tomograms with 99 Tc-HM-PAO in patients with dementia (Alzheimer type and HIV) and Parkinson's disease – Initial results. J Cereb Blood Flow Metab 8: 109–115

De Jong BM, Van Royen EA (1989) Uptake of SPECT radiopharmaceuticals in neocortical brain cultures. Eur J Nucl Med 15: 16–20

Delvenne V, Delecluse F, Hubain Ph, et al (1990) Regional blood flow in patients with affective disorders. Br J Psychiatry 157: 359–365

Frackowiak RSJ, Gibbs JM (1983) Cerebral metabolism and blood flow in normal and pathological aging. In: PL Magistretti (ed) Functional Radionuclide Imaging of the Brain. Raven Press, New York, pp 305–309

Gibb WRG (1994) Cortical dementia in Parkinson's disease. In: Wolters EC, Scheltens P (eds) Mental dysfunction in Parkinson's disease. Current issues in neurodegenerative disorders, vol 1. ICG Publications, Dordrecht, pp 211–221

Globus M, Mildworf B, Melamed E (1985) Cerebral blood flow and cognitive impairment in Parkinson's disease. Neurology 35: 1135–1139

Gur RE, Skolnick BE, Gur RC (1984) Brain functions in psychiatric disorders. II. Regional cerebral blood flow in medicated unipolar depressives. Arch Gen Psychiatry 41: 695–699

Henriksen L, Boas J (1985) Regional cerebral blood flow in hemiparkinson patients. Emission computerized tomography of inaled 133-Xenon before and after levodopa. Acta Neurol Scand 71: 257–266

Jagust WJ, Reed BR, Martin EM, Eberling JL, Nelson-Abbott RA (1992) Cognitive function and regional cerebral blood flow in Parkinson's disease. Brain 115: 521–537

Jolles JJ (1994) Brain and behavior mechanisms in depression. In: Wolters EC, Scheltens P (eds) Mental dysfunction in Parkinson's disease. Current issues in neurodegenerative disorders, vol 1. ICG Publications, Dordrecht, pp 303–315

Kuhl DE, Metter EJ, Riege WH (1984) Patterns of local cerebral glucose utilization determined in Parkinson's disease by the 18F fluorodeoxyglucose method. Ann Neurol 15: 419–424

Kuhl DE, Metter EJ, Benson DF, et al (1985) Similarities of cerebral glucose metabolism in Alzheimer's and parkinsonian dementia. J Cereb Blood Flow Metab 5 [Suppl 1]: S169

Kuiper MA, Scheltens P, Weinstein HC, Wolters EC (1993) Dementia in Parkinson's disease compared with Alzheimer's type dementia. A [99mTc]HMPAO SPECT study. In: Nicolini M, Zatta PF, Corain B (eds) Alzheimer's disease and related disorders. Pergamon Press, Oxford, pp 45–48

Kuiper MA, Weinstein HC, Bergmans PLM, Scheltens P, Wolters EC (1992) [99mTc] HMPAO SPECT and dementia in Parkinson's disease. J Neurol Neurosurg Psychiatry 55: 981

Kuniyoshi M, Arikawa K, Miura C, Inanaga K (1989) Parkinsonism manifesting with depression as the first sign. Jpn J Psychiatry Neurol 43: 37–43

Lenzi GL, Jones T, Reid GL, Moss S (1979) Regional impairment of cerebral oxidative metabolism in Parkinson's disease. J Neurol Neurosurg Psychiatry 22: 59–62

Lipsey JR, Robinson RG, Pearlson GD, et al (1983) Mood change following bilateral hemisphere brain injury. Br J Psychiatry 143: 266–273

Lishman WA (1968) Brain damage in relation to psychiatric disability after head injury. Br J Psychiatry 144: 373–410

Liu RS, Lin KN, Wang SJ, et al (1992) Cognition and [99mTc]HMPAO SPECT in Parkinson's disease. Nucl Med Commun 13: 744–748

Maes M, Dierckx R, Meltzer HY, et al (1993) Regional cerebral blood flow in unipolar depression measured with Tc-99m-HMPAO single photon emission computed tomography: negative findings. Psychiatry Res 50: 77–88

Martin WRW, Beckman JH, Calne DB, et al (1984) Cerebral glucose metabolism in Parkinson's disease. Can J Neurol Sci 11: 169–173

Mayberg HS, Jeffery PJ, Wagner HN, Simpson SG (1991) Regional cerebral blood flow in patients with refractory unipolar depression measured with Tc-99-HMPAO SPECT. J Nucl Med 32: S951

Mayberg HS, Starkstein SE, Sadzot B, et al (1990) Selective hypometabolism in the inferior frontal lobe in depressed patients with Parkinson's disease. Ann Neurol 28: 57–64

Neumann C, Baas H, Hefner R, Hör G (1989) SPECT-Befunde bei Hemiparkinson-Syndrom mit 99m Tc-HMPAO. Nucl Med 28: 92–94

Pizzolato G, Dam M, Borsato N, et al (1988) [^{99m}Tc]HM-PAO SPECT in Parkinson's disease. J Cereb Blood Flow Metab 8: S101–S108

Sackheim HA, Greenberg MS, Weinan Al, et al (1982) Hemispheric assymetry in the expression of positive and negative emotions. Arch Neurol 39: 210–218

Sawada H, Udaka F, Kameyama M, et al (1992) SPECT findings in Parkinson's disease associated with dementia. J Neurol Neurosurg Psychiatry 55: 960–963

Spampinato U, Habert MO, Mas JL, et al (1991) [^{99m}Tc]HM-PAO SPECT and cognitive impairment in Parkinson's disease: a comparison with dementia of the Alzheimer type. J Neurol Neurosurg Psychiatry 54: 787–792

Sperling R, Becker KA, Satlin A, Garada B, Holman BL, Growdon JH (1993) SPECT cerebral perfusion in neurodegenerative disease with dementia. Neurology 43 [Suppl]: A406

Tachibana H, Kawabata K, Tomino M, et al (1993) Brain perfusion imaging in Parkinson's disease and Alzheimer's disease demonstrated by three-demensional surface display with 123I-iodoamphetamine. Dementia 4: 334–341

Tröster AI, Paolo AM, Lyons KE, Glatt SL, Hubble JP, Koller WC (1995) The influence of depression on cognition in Parkinson's disease. Neurology 45: 672–676

Trzepac PT, Hertweck M, Starratt C, Zimmerman L, Adatepe MH (1992) The relationship of SPECT scans to behavioral dysfunction in neuropsychiatric patients. Psychosomatics 33: 62–71

Uytenhoef P, Portelange P, Jacquy J, et al (1983) Regional cerebral blood flow and lateralized hemispheric dysfunction in depression. Br J Psychiatry 143: 128–132

Van Praag HM (1994) Serotonin and affective psychopathology in parkinson's disease. A psychobiological hypothesis with therapeutic consequences. In: Wolters EC, Scheltens P (eds) Mental dysfunction in Parkinson's disease. Current issues in neurodegenerative disorders, vol 1. ICG Publications, Dordrecht, pp 335–351

Wang S-J, Liu R-S, Liu H-S, et al (1993) Technetium-99m hexamethylpropylene amine oxime single photon emission tomography of the brain in early Parkinson's disease: correlation with dementia and lateralization. Eur J Nucl Med 20: 339–344

Weinstein HC, Scheltens P, Van Royen EA, Hydra A (1993) Neuroimaging in the diagnosis of Alzheimer's disease. II. Positron and single photon emission tomography (PET and SPECT). Clin Neurol Neurosurg 95: 81–91

Authors' address: Dr. S. Bissessur, Department of Neurology, Academisch Ziekenhuis VU Amsterdam, P.O. Box 7057, 1007 MB Amsterdam, The Netherlands.

IBZM- and CIT-SPECT of the dopaminergic system in parkinsonism

G. Tissingh[1], J. Booij[2], A. Winogrodzka[1], E. A. van Royen[2], and E. Ch. Wolters[1]

[1]Graduate School Neurosciences Amsterdam, Department of Neurology, Academisch Ziekenhuis VU Amsterdam, [2]Department of Nuclear Medicine, Academic Medical Center University of Amsterdam, Amsterdam, The Netherlands

Summary. Parkinsonism is most of the time caused by idiopathic Parkinson's disease (IPD). Considering the differences in therapeutic response and prognosis, in vivo discrimination between IPD and "parkinsonism-plus" syndromes is important. Recently, ligands have become available for imaging the pre- and postsynaptic dopaminergic system by Single Photon Emission Computed Tomography (SPECT). Visualization of postsynaptic D_2 dopamine receptors using ^{123}I-iodobenzamide (^{123}I-IBZM) may contribute to the differential diagnosis between IPD and "parkinsonism-plus" syndromes as IPD is a pure presynaptic disease. Imaging of the presynaptic dopamine transporters using $[^{123}I]\beta$-CIT (2β-carbomethoxy-3β-(4-iodophenyl)tropane) may be used as a diagnostic technique. Early disease detection in subjects suspected to be at risk for developing IPD has become possible using $[^{123}I]\beta$-CIT or other ligands for the dopamine transporter. Furthermore, with SPECT one is probably able to monitor in an objective way the efficacy of new pharmacological therapies.

Introduction

Idiopathic Parkinson's disease (IPD) is by far the most common cause of parkinsonism (Rajput et al., 1984). The predominant neuropathological feature in IPD is a degeneration of the dopaminergic cells in the substantia nigra, pars compacta, and the ventral tegmental area. This results in a marked loss of striatal dopamine (DA) (Bernheimer et al., 1973). Striatal DA concentrations must be significantly reduced before clinical symptoms will become apparent. The main symptoms are bradykinesia, tremor, rigidity and postural instability. A good response to dopaminergic drugs, i.e. levodopa and DA receptor agonists, is commonly considered to be one of the hallmarks of IPD. Another disorder that causes parkinsonism is multiple system atrophy (MSA), which may present with any combination of extrapyramidal, pyramidal, cerebellar and autonomic features. Two types can be distinguished: the striato-nigral type with predominantly extrapyramidal signs, and the olivopontocerebellar type with mainly cerebellar features (Quinn, 1989).

Neuropathologically, MSA is characterized by typical oligodendroglial and intraneuronal cytoplasmatic inclusions, and cell loss and gliosis in at least 2 of the following structures: striatum, substantia nigra, locus coeruleus, inferior olives, cerebellar Purkinje cells, pontine nuclei, middle cerebellar peduncles, and intermediolateral cell columns (Quinn, 1989). In contrast to IPD, patients with MSA usually show a lack of response to dopaminergic medication. Corticobasal ganglionic degeneration and the Steel-Richardson-Olszewski or progressive supranuclear palsy (SRO/PSP) syndrome are also neurodegenerative disorders, with signs and symptoms of parkinsonism with a transient or poor response to levodopa. Together with MSA, they are the so-called "parkinsonism-plus" syndromes, and they may account for up to 10-15% of the patients with parkinsonism (Quinn, 1989). As IPD is a purely presynaptic disease (whereas the 'parkinsonism-plus' syndromes display pre- as well as postsynaptic abnormalities) functional imaging of the brain with ligands for the dopaminergic system may be useful in the (early) diagnosis of IPD. Recently, these ligands have become available for imaging the pre- and postsynaptic dopaminergic system in vivo by positron emission tomography (PET) and single photon emission computed tomography (SPECT) (Brooks, 1993; Innis, 1994). Application of these techniques may increase the accuracy of the clinical diagnosis of IPD, which is estimated to be approximately 80% (Hughes et al, 1992), by differentiating in vivo, between IPD and the 'parkinsonism-plus' syndromes. Considering the differences in therapeutic response and prognosis, differential diagnosis between IPD and other parkinsonian syndromes is important in patient care. This chapter will briefly review the use of SPECT in patients with parkinsonism.

Principles of SPECT

In SPECT research radio-labeled, gamma X-rays emitting radionuclides are used. Most SPECT radioligands are labeled with [123]Iodine, which is highly suitable for labeling many molecules. It's half life is 13 hours and this results in an acceptable radiation dose to the patient and a suitable energy of 159 keV for imaging with the gamma camera. The local distribution of the SPECT ligands can be visualized by the use of a rotating or a multidetector camera system. In a rotating camera system the camera rotates around the head and records two-dimensional images from different angles. A multidetector camera contains paired detectors, arranged opposite each other. After detection of the radioactivity, the tomogram is reconstructed by a computer that monitors precisely and immediately the amount of activity at a certain location, combining the data of all detectors. Most of the time a semi-quantitative method is used for analysis of the SPECT images by comparing the specific to non-specific binding in the regions of interest (for review brain SPECT, see Verhoeff et al., 1992 and Rosenthal et al., 1995).

An absolute quantification of the radiolabeled ligand is a problem in SPECT, due to the physical characteristics of gamma radiation which is absorbed by tissues. However, this can be corrected to a certain degree by the use of mathematical attenuation correction models. With the introduction of digital image matching with computed tomography (CT) or magnetic resonance imaging (MRI), the quantitative power of SPECT may be much improved in the future.

Postsynaptic dopaminergic system

With the introduction of radioligands with a high selectivity for the D_2 DA receptor, in vivo assessment of the postsynaptic D_2 receptor density in the human brain has become possible (Kung et al., 1990). It has been shown that these receptors may be visualized and quantified reliably using ^{123}I-iodobenzamide (^{123}I-IBZM) SPECT (Brücke et al., 1991). Studies using this imaging technique are useful for the differentiation between various disorders of the dopaminergic system (Tatsch et al., 1991). The postsynaptic D_2 DA receptor is the main target for antiparkinsonian medication. Post mortem studies have shown that the absence of an adequate response to dopaminergic medication in parkinsonism may be caused by a decreased postsynaptic D_2 receptor density in the striatum (Pascual et al., 1992; Churchyard et al., 1993). A reduced number of these receptors may be taken as an indicator of degeneration of striatal interneurons and neurons projecting to the external pallidal segment (Harrison et al., 1992). Several ^{123}I-IBZM SPECT studies are in agreement with the post mortem observations, finding a reduced striatal D_2 uptake in patients with MSA and SRO (Van Royen et al., 1993; Schulz et al., 1994).

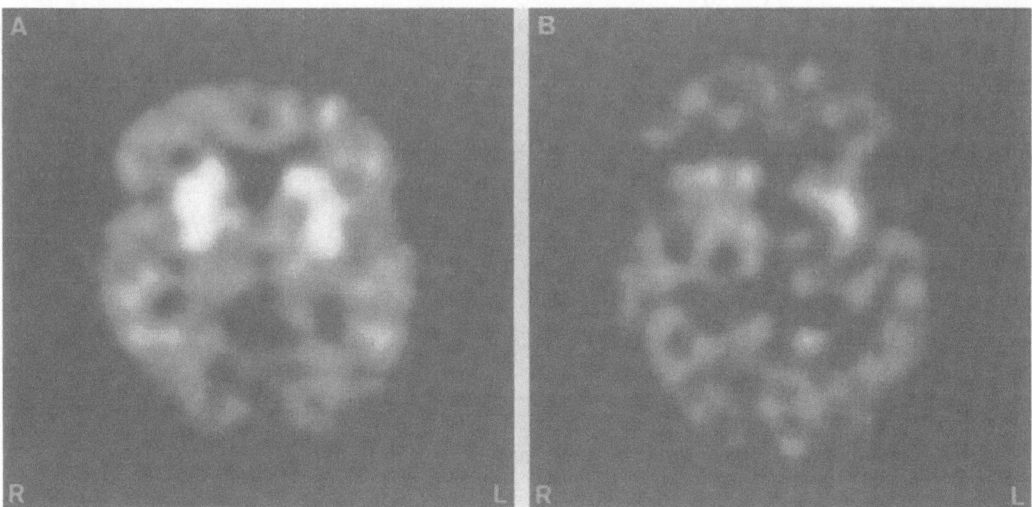

Fig. 1. [123I]IBZM SPECT images of a healthy volunteer (**A**) and a patient with MSA (**B**). Transverse slices from the brain at the level of the striatum (*L* left, *R* right). In both images the level of [123I]IBZM activity is colour encoded from low (black) to high (white), and scaled (corrected for the injected dose/kg) to the maximum in the slice of the control

Subjects with IPD show normal striatal D_2 binding, reflecting a normal integrity of postsynaptic D_2 DA receptors in the striatum (Brücke et al., 1991; Laulumaa et al., 1993; Schwarz et al., 1993). Thus, IBZM-SPECT may contribute to the differential diagnosis between MSA and SRO, and IPD. Visualization of postsynaptic D_2 DA receptors may also help to predict the response to dopaminergic medication. Apomorphine testing has proved to be a reliable and quick way to predict the dopaminergic responsiveness (Oertel et al., 1989; Hughes et al., 1990). In a prospective study it was found that IBZM-SPECT results predicted a positive or negative response to apomorphine in 30 out of 34 patients, and response to dopaminergic therapy in 27 out of 31 subjects with de novo parkinsonism (Schwarz et al., 1992). Together with a positive apomorphine test, normal IBZM binding strongly supports the clinical diagnosis of IPD (Oertel et al., 1993).

Presynaptic dopaminergic system

The dopaminergic cell loss in IPD is accompanied by a decrease in the striatal DA transporter, located on the nerve terminals of dopaminergic neurons projecting from the substantia nigra to the striatum (Kaufman and Madras, 1991). Recently, several ligands derived from cocaine have been developed for SPECT imaging of these DA transporters (Neumeyer et al., 1991, 1994; Carrol et al., 1995). One of these compounds, [^{123}I]β-CIT (2β-carbomethoxy-3β-(4-iodophenyl)tropane), has proved to be especially useful in monitoring the dopaminergic degeneration of the nigrostriatal pathway in IPD patients. Although [^{123}I]β-CIT binds to serotonin transporters with approximately equivalent affinity as to DA transporters, Laruelle and co-workers showed in primate studies that the striatal [^{123}I]β-CIT binding is associated with the DA transporter (Laruelle et al., 1993). This finding is consistent with the known relative density of DA to serotonin transporters in striatum of > 10:1 (Innis, 1994).

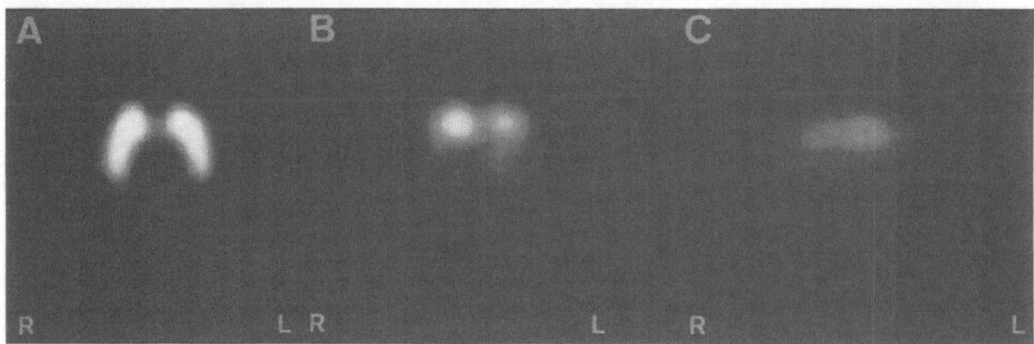

Fig. 2. [^{123}I]β-CIT SPECT images of a healthy volunteer (**A**), an early IPD patient (**B**), and a late IPD patient (**C**). Transverse slices from the brain at the level of the striatum (*L* left, *R* right). In all images the level of [^{123}I]β-CIT activity is colour encoded from low (black) to high (white), and scaled (corrected for the injected dose/kg) to the maximum in the slice of the control

Initial SPECT studies imaging the DA transporter with $[^{123}I]\beta$-CIT showed severe loss of striatal DA transporters in IPD compared with healthy human subjects (Innis et al., 1993; Brücke et al., 1993). This finding has been confirmed in more recent studies (Seibyl et al., 1995; Vermeulen et al., 1995; Rinne et al., 1995), showing markedly abnormal striatal uptake, more pronounced in the putamen than in the caudate nucleus.

This is in agreement with the finding of initial loss of DA in the putamen in IPD patients (Kish et al., 1988). It appeared that there were greater losses in the striatum on the side contralateral to where the parkinsonian signs started. Also, decreased striatal uptake correlates well with symptom severity (Seibyl et al., 1995; Rinne et al., 1995). Similar to IPD patients, subjects with MSA show loss of DA transporters in the striatum, more dramatically in the putamen than in the caudate nucleus (Booij et al., 1995). This correlates with pathological studies which showed that in MSA the caudate nucleus is less seriously affected than the putamen (Spokes et al., 1979; Fearnley and Lees, 1990). However, in contrast to this it has been suggested that patients with MSA show a more uniform and symmetrical loss of dopaminergic activity, both involving the caudate and putamen, compared to patients with IPD (Brooks et al., 1990a; Innis, 1994). A PET study showed a uniform striatal involvement in some patients, as well as a relative sparing of ^{18}F-DOPA uptake in the caudate nucleus of the other examined MSA patients (Brooks et al., 1990b). So, it is probable that $[^{123}I]\beta$-CIT only allows us to differentiate between patients with IPD and MSA with preferential involvement of the putamen on the one hand, and MSA patients in wich the putamen is just as much affected as the caudate nucleus on the other hand.

Conclusion

SPECT imaging of the dopaminergic system offers many possibilities. IBZM-SPECT is useful in differentiating between various parkinsonian syndromes and it may help to predict the dopaminergic response. Moreover, visualizing the D_2 receptor can also be used to form homogeneous patient populations for clinical trials, studying both experimental pharmacological and surgical therapies. Imaging the DA transporter may be used as a diagnostic technique, even during the preclinical phase of IPD, which may be as long as 20 to 40 years. Early disease detection in subjects suspected to be at risk of developing IPD (e.g. based on neuropsychological and clinical features) creates the chance for putative neuroprotective, pharmacological treatment which may slow down the progression of dopaminergic cell loss. Furthermore, it is probable that SPECT will provide an objective and reliable way to monitor the efficacy of these new neuropharmacological therapies.

References

Bernheimer H, Birkmayer W, Hornykiewicz O, Jellinger K, Seitelberger F (1973) Brain Dopamine and the syndromes of Parkinson and Huntington. Clinical, morphological and neurochemical correlations. J Neurol Sci 20: 415–455

Booij J, Vermeulen RJ, Tissingh G, et al (1995) Visualization of the dopamine transporter in Parkinson's disease and multiple system atrophy: a [123-I]β-CIT study. Acta Neurol Belg 95 [Suppl]: 35

Brooks DJ, Ibanez V, Sawle GV, et al (1990a) Differing patterns of striatal ^{18}F-DOPA uptake in Parkinson's disease, multiple system atrophy and progressive supranuclear palsy. Ann Neurol 28: 547–555

Brooks DJ, Salmon EP, Mathias CJ, et al (1990b) The relationship between locomotor disability, autonomic dysfunction, and the integrity of the striatal dopaminergic system in patients with multiple system atrophy, pure autonomic failure, and Parkinson's disease, studied with PET. Brain 113: 1539–1552

Brooks DJ (1993) Functional imaging in relation to parkinsonian syndromes. J Neurol Sci 115: 1–17

Brücke T, Podreka I, Angelberger P, et al (1991) Dopamine D$_2$ receptor imaging with SPECT: studies in different neuropsychiatric disorders. J Cereb Blood Flow Metab 11: 220–228

Brücke T, Wenger S, Asenbaum S, et al (1993) Dopamine D$_2$ receptor imaging and measurement with SPECT. In: Narabayashi H, Nagatsu T, Yanagisawa N, Mizuno Y (eds) Advances in Neurology, vol 60. Raven Press, New York

Carroll FI, Scheffel U, Dannals RF, Boja JW, Kuhar MJ (1995) Development of imaging agents for the dopamine transporter. Med Res Rev 15: 419–444

Churchyard A, Donnan GA, Hughes A, et al (1993) Dopa resistance in multiple-system atrophy: loss of postsynaptic D$_2$ receptors. Ann Neurol 34: 219–226

Fearnley JM, Lees AJ (1990) Striatonigral degeneration: a clinico-pathological study. Brain 113: 1823–1842

Harrison MB, Wiley RG, Wooten GF (1992) Changes in D$_2$ but not D$_1$ receptor binding in the striatum following selective lesion of striatopallidal neurons. Brain Res 590: 305–310

Hughes A, Lees AJ, Stern GM (1990) Apomorphine test to predict dopaminergic responsiveness in parkinsonian syndromes. Lancet 336: 32–34

Hughes AJ, Daniel SE, Kilford L, Lees AJ (1992) Accuracy of clinical diagnosis of idiopathic Parkinson's disease: a clinico-pathological study of 100 cases. J Neurol Neurosurg Psychiatry 55: 181–184

Innis RB, Seibyl JP, Scanley BE, et al (1993) Single photon emission computed tomographic imaging demonstrates loss of striatal dopamine transporters in Parkinson's disease. Proc Natl Acad Sci USA 90: 11965–11969

Innis RB (1994) Single-photon emission tomography imaging of dopamine terminal innervation: a potential clinical tool in Parkinson's disease. Eur J Nucl Med 21: 1–5

Kaufman MJ, Madras BK (1991) Severe depletion of cocaine recognition sites associated with the dopamine transporter in Parkinson's disease striatum. Synapse 9: 43–49

Kish SJ, Shannak K, Hornykiewicz O (1988) Uneven pattern of dopamine loss in the striatum of patients with idiopathic Parkinson's disease. N Engl J Med 318: 876–880

Kung HF, Alavi A, Chang W, et al (1990) In vivo SPECT imaging of CNS D-2 dopamine receptors: initial studies with iodine-123-IBZM in humans. J Nucl Med 31: 573–579

Laruelle M, Baldwin RM, Malison RT, et al (1993) SPECT imaging of dopamine and serotonin transporters with [^{123}I]β-CIT: pharmacological characterization of brain uptake in nonhuman primates. Synapse 13: 295–309

Laulumaa V, Kuikka JT, Soininen H, Bergström K, Länsimies E, Riekkinen P (1993) Imaging of D$_2$ dopamine receptors of patients with Parkinson's disease using single photon emission computed tomography and iodobenzamide I 123. Arch Neurol 50: 509–512

Neumeyer JL, Wang S, Millius RA, et al (1991) [^{123}I]-2β-carbomethoxy-3β-(4- iodophenyl) tropane: high-affinity SPECT radiotracer of monoamine reuptake sites in brain. J Med Chem 34: 3144–3146

Neumeyer JL, Wang S, Gao Y, et al (1994) N-3-fluoroalkyl analogs of (1R)-2β-carbomethoxy-3β-(4-iodophenyl)-tropane (β-CIT) radiotracers for positron emission tomography and single photon emission computed tomography imaging of dopamine transporters. J Med Chem 37: 1558–1561

Oertel WH, Gasser T, Ippisch R, Trenkwalder C, Poewe W (1989) Apomorphine test for dopaminergic responsiveness. Lancet 1: 1262–1263

Oertel WH, Schwarz J, Tatsch K, Arnold G, Gasser T, Kirsch C-M (1993) IBZM-SPECT as predictor for dopamimetic responsiveness of patients with de novo parkinsonian syndrome. In: Narabayashi H, Nagatsu T, Yanagisawa N, Mizuno Y (eds) Advances in neurology, vol 60. Raven Press, New York

Pascual J, Berciano J, Grijalba B, Del Olmo E, González AM, Figols J, Pazos A (1992) Dopamine D_1 and D_2 receptors in progressive supranuclear palsy: an autoradiographic study. Ann Neurol 32: 703–707

Quinn N (1989) Multiple system atrophy – the nature of the beast. J Neurol Neurosurg Psychiatry [spec Suppl] 78–89

Rajput AH, Offord KP, Beard CM, Kurland LT (1984) Epidemiology of parkinsonism: incidence, classification, and mortality. Ann Neurol 16: 278–282

Rinne JO, Kuikka JT, Bergström KA, Rinne UK (1995) Striatal dopamine transporter in different disability stages of Parkinson's disease studied with [123I]β-CIT SPECT. Parkinsonism & Related Disorders 1: 47–51

Rosenthal MS, Cullom J, Hawkins W, Moore SC, Tsui BMW, Yester M (1995) Quantitative SPECT imaging: a review and recommendations by the focus committee of the society of nuclear medicine computer and instrumentation council. J Nucl Med 36: 1489–1513

Schulz JB, Klockgether T, Petersen D, et al (1994) Multiple system atrophy: natural history, MRI morphology, and dopamine receptor imaging with 123IBZM-SPECT. J Neurol Neurosurg Psychiatry 57: 1047–1056.

Schwarz J, Tatsch K, Arnold G, Gasser T, Trenkwalder C, Kirsch CM, Oertel WH (1992) 123I-Iodobenzamide-SPECT predicts dopaminergic responsiveness in patients with de novo parkinsonism. Neurology 42: 556–561

Schwarz J, Tatsch K, Arnold G, Ott M, Trenkwalder C, Kirsch CM, Oertel WH (1993) 123I-Iodobenzamide-SPECT in 83 patients with de novo parkinsonism. Neurology 43 [Suppl 6]: S17–20

Seibyl JP, Marek KL, Quinlan D, et al (1995) Decreased single-photon emission computed tomographic [123I]β-CIT striatal uptake correlates with symptom severity in Parkinson's disease. Ann Neurol 38: 589–598

Spokes EGS, Bannister R, Oppenheimer DR (1979) Multiple system atrophy with autonomic failure. Clinical, histological, and neurochemical observations on four cases. J Neurol Sci 43: 59–82

Tatsch K, Schwarz J, Oertel WH, Kirsch C-M (1991) SPECT imaging of dopamine D_2 receptors with 123I-IBZM: initial experience in controls and patients with Parkinson's syndrome and Wilson's disease. Nucl Med Commun 12: 699–707

Van Royen E, Verhoeff NFLG, Speelman JD, Wolters EC, Kuiper MA, Janssen AGM (1993) Multiple system atrophy and progressive supranuclear palsy. Diminished striatal D_2 dopamine receptor activity demonstrated by 123I-IBZM single photon emission computed tomography. Arch Neurol 50: 513–516

Verhoeff NPLG, Buell U, Costa DC, et al (1992) Basics and recommendations for brain SPECT. Nuklearmedizin 31: 114–131

Vermeulen RJ, Wolters ECh, Tissingh G, et al (1995) Evaluation of [123I]β-CIT binding with SPECT in controls, early and late Parkinson's disease. Nucl Med Biol 22: 985–991

Authors' address: Dr. G. Tissingh, Department of Neurology, Academisch Ziekenhuis VU Amsterdam, P.O. Box 7057, 1007 MB Amsterdam, The Netherlands.

Pathophysiology of movement disorders studied using PET

K. L. Leenders

Paul Scherrer Institute, Villigen, Switzerland

Summary. PET radiotracer methods can measure various biochemical features of brain tissue in the living human brain. Here, local brain energy consumption and striatal dopaminergic function will be discussed in the light of the neurodegenerative processes underlying Parkinson's disease. Particularly, disease progression and its consequences for protective and restorative strategies will be outlined. Also, an example will be given to demonstrate how the effect of neurotrophic factors on the striatal dopaminergic system can be monitored by PET tracer methods.

Introduction

In this chapter examples will be given of how PET methods can help unravel pathophysiology of movement disorders like Parkinson's disease (PD) by investigating biochemical features of the living patient's brain. The direct study of the striatal dopaminergic status will mainly be discussed here. Particularly the advantages of studying disease progression and the resulting options for neuroprotection or neurorestoration will be outlined.

Only briefly reviewed here are the recently increasing possibilities of investigating brain energy metabolism using modern pattern or covariance analysis methods of cerebral glucose consumption. The PET activation studies concerning cerebral motor organisation or cognition are outside the scope of this chapter, although the impact of these studies on the improvement of our understanding of pathophysiology of movement disorders must be stressed. It can be expected that in the near future much information will be obtained by these methods.

Methodological aspects

Using radiolabelled tracer substances in vivo regional cerebral biochemical activity can be measured by positron emission tomography (PET) and to a certain extent also by single photon emission computer tomography (SPECT). Applying PET radiotracers, brain energy metabolism or receptor binding or blood-brain barrier transport can be quantitated in absolute units. PET uses tracer substances which are physiological or near-physiological, e.g. water labelled with O-15 or fluoro-deoxyglucose (FDG) labelled with F-18. The first has a radioactive decay half-life of 2 minutes and the latter of two hours. Tracers which measure features of the cerebral dopamine neurotransmitter system are e.g. F-18 labelled fluoro-L-dopa (FDOPA) and C-11 labelled raclopride. The first tracer yields a measure for pre-synaptic dopadecarboxy-lase capacity of the dopaminergic nigrostriatal nerve terminals in the striatum. The latter indicates D2 receptor density which in striatum is mainly located post-synaptically on projection neurons.

Brain glucose metabolism

Regional energy consumption in rest is mainly determined by local synaptic density and not so much by the local number of neurons. It is in the nerve terminals that mitochondrial concentration is high. Local changes of energy consumption in the brain should in the first instance be viewed as the effect of changed activity or density of the afferent nerve terminals into that region. Loss of neurons in one place may show itself as reduced energy consumption at a distance provided the projection of the lost neuronal connections is intense enough to be measurable. The reverse, increased local energy consumption by increased activation of afferent neurons elsewhere, can also occur. This princi-ple has been documented using autoradiographic deoxyglucose methods.

Global cerebral metabolism in PD patients is on average moderately or even markedly reduced (20% or more) compared to control values also in early PD as was demonstrated by several authors using PET methods (Kuhl et al., 1984 and others; for review see Leenders, 1990; Eidelberg et al., 1995a). The cause for this is not clear, although an obvious reason could be a general loss of cerebral neurons in the PD brain. This however has not been demon-strated yet. An alternative mechanism could be general deafferentation of the whole brain as a result of the impaired dopaminergic (or catecholaminergic) input from the brain stem.

On the other hand the loss of dopaminergic input in PD from the substantia nigra into striatal regions (and possibly also into cortical regions) leads to a characteristic pattern of regional energy metabolism changes within the large hemispheres due to alterations of neuronal activities in the various chaines of nerve cell interconnections. A key-role in PD apparently plays the increased pallidothalamic output on the basis of a disturbed balance between the direct and indirect striato-palidal pathways. This in turn is to result in increased tha-lamic inhibition and concurrently in increased metabolic activity in the pallido-thalamic target regions which has been shown by various authors (Eidelberg et al., 1994; Marie et al., 1995). It is further supposed that in PD the reduced thalamo-cortical glutamatergic projections to the prefrontal cortex are respon-sible for the decrease of metabolism in that region. Prefrontal cortex is thought to play an important role as a central motor executive. Impaired function of this region is associated with diminished performance in various frontal lobe neu-ropsychological function tests of PD patients.

First efforts to unravel specific topographic cerebral metabolic patterns in PD concordant with the pattern of brain regions involved in motor control have been made using regional metabolic glucose consumption PET measurements and covariance analyses (Eidelberg et al., 1994, 1995a, 1995b). It was shown that in PD a disease-specific regional covariance pattern characterized by bilateral lentiform and thalamic hypermetabolism and premotor hypometabolism was present. This statistical analysis allows to assign a metabolic profile score to each subject and compare that with clinical items such as disease severity.

Disease progression

The disorganisation in PD of the motor (and other) functions of the large hemispheres of the brain, particularly the frontal lobe, stems from the failure to modulate striatal transmission by a rather amazingly small number of dopamine neurons in the substantia nigra. The cause of the degeneration of the dopamine neurons in PD is not known and not all dopamine neurons die. It has been unknown until recently how slow or how fast the loss of these neurons in general is and which part of the dopamine system remains. It is common praxis to state that at least 80% of the dopamine cells in the substantia nigra need to be lost before parkinsonian signs and symptoms become manifest. However, that is unlikely to be true. Indeed, the endogenous dopamine concentration in striatum or substantia nigra will be very low indeed, a few percent of control values, however the cellular structures, i.e. the nerve cell body in the substantia nigra plus the associated nerve terminals in the striatum, are likely to be present in much higher percentage. Whatever the cause of PD, first the endogenous dopamine production is affected and only later the neurons themselves are lost. This is an important distinction, since relative or total preservation of cells for some time e.g. several years after onset of clinical signs, will provide the basis for possible neuroprotective therapeutic strategies (see below). It also explains why exogenous levodopa is so effective particularly in the early stages of PD: enough dopaminergic striatal nerve terminals are left (with very low endogenous dopamine production) to take up the exogenous levodopa and decarboxylate it into dopamine. Dopaminergic nerve terminals are very efficient in taking up extracellular dopa or dopamine, to keep it contained and use it locally where it is needed for signal modulation. Thus early on in the disease relatively little exogenous levodopa is usually needed to provide effective treatment. Also the duration of the treatment effect after one oral dose lasts longer than in later stages of the disease. It is assumed here that this is due to slow reduction of dopaminergic nerve terminals in the striatum during disease development. In PD the post-synaptic dopaminergic receptor system is intact (no loss of striatal projection neurons) further guaranteeing that complete or partial replenishing of striatal nerve terminals with dopamine restores impaired neurotransmisson.

In the light of above argumentation it will be necessary to determine how exacly the time course of striatal nerve terminal loss in PD is. Using PET tracers specifically measuring biochemical features of these nerve terminals it is possible to follow the progression of decline in living PD patients. The tracer fluorodopa (FDOPA) labelled with F-18 measures dopadecarboxylase, which in striatum reflects in general the volume of dopaminergic nerve terminals. Other tracers may become increasingly used like the cocaine derived substances e.g. CIT related tracers, which indicate pre-synaptic dopaminergic dopamine reuptake sites and thus are also a measure of dopaminergic nerve terminals. Here, we consider only FDOPA results. Various groups have demonstrated that specific FDOPA uptake in putamen of PD patients is on average approximately 40% of control values and in caudate nucleus 60% (see e.g. Leenders et al., 1990). In early PD these values can be considerably higher and for caudate nucleus will often be normal. Cross-sectional analysis suggests that the decline of dopadecarboxylase capacity has a fairly rapid phase during the first three years after onset of clinical signs and after that shows a plateauing almost always leaving 10 to 20% of normal capacity.

An illustration of these findings is given in Fig. 1 for putamen. The results suggest that the dopaminergic decline in putamen of PD patients does not start very long before the clinical signs and symptoms become clearly manifest. Extrapolation of the available cross-sectional FDOPA PET results indicate approximately 4 years. This would be in line with post-mortem data (Fearnly and Lees, 1991) and other PET results (Morrish et al., 1995). However, longitudinal data will be necessary to confirm this. It needs to be kept in mind that intact or only slightly reduced FDOPA uptake in striatum reflects dopadecarboxylase activity of the dopamine nerve terminals. Thus it could well be that other biochemical indicators of the nigro-striatal neurotransmitter status, e.g. endogenous dopamine concentration itself, may be changed from a much earlier time point. This might explain why certain subjects have noticed subtle phenomena which were present long before the actual disease became

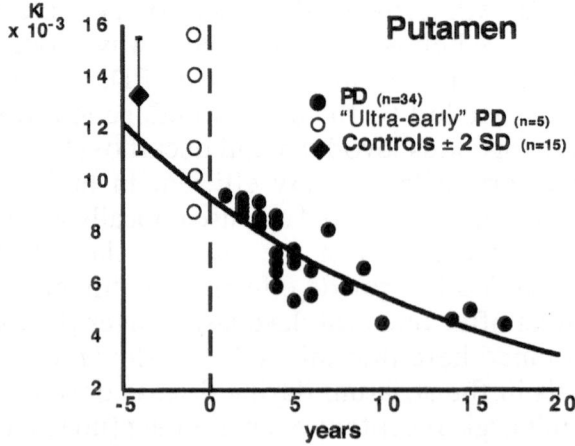

Fig 1. Specific FDOPA uptake in putamen versus disease duration in years in PD patients. Specific FDOPA uptake is expressed as influx constant Ki (unit: 1/min times 10^{-3}) according to Patlak and Blasberg (1985). The error bars of the control values comprise two standard deviations on each side. For description of the "ultra-early" PD patients see text

clearly manifest. In any event it appears that even after the disease has become clinically manifest there remains a large body of nerve cells to be protected from neurodegenerative decline if protective strategies become effective. Even if under protective therapy the number of dopamine nerve cells were below normal it could be expected that sufficient cells would remain to let the standard antiparkinsonian drugs be effective over much longer periods of time than now is common.

Also, it is not clear yet whether the decline is linear or rather exponential. This has some significance since according to the mathematical expression an exponential decline would suggest loss of a constant fraction over time. This in turn could be interpreted to mean that once the disease process has started the pathological drive remains constant.

In addition to the 34 PD patients plotted in Fig. 1, also specific FDOPA uptake indicated by the influx constant Ki (Patlak and Blasberg, 1985) of five subjects who possibly were in a very early phase of PD, is given. The selection criteria are described elsewhere (de Rijk et al., 1995). It will be important to follow these subjects up to determine whether they indeed develop classical PD and how their FDOPA uptake changes accordingly. It would reinforce some of the above mentioned hypotheses.

Neuronal protection and restoration

Drugs

Several drugs have been suggested to possess neuroprotective properties and to halt or delay further decline of dopamine neurons. To prove such phenomena in PD patients on pure clinical grounds is difficult because of the slow progression of the disease and the various ways symptoms may present themselves. Large groups of patients need to be followed over long periods with all the inherent practical and financial consequences. If under such circumstances symptomatic drug effects need to be disentangled from a possible small protective effect, the interpretation then becomes difficult or impossible. The recent large selegiline study illustrates these difficulties.

An improvement to test substances for the above purpose might be when the status of the striatal dopaminergic system is taken into account using PET tracer methods and directly compared with associated clinical changes. Such a study is now under way testing pergolide monotherapy versus levodopa monotherapy in 100 patients over a period of three years. The results of this and other studies are eagerly awaited.

Neurotrophic factors

Neurotrophic factors (NTF) have become of increasing interest in recent years for their potential to protect or restore neurons in various experimental conditions (Tomac et al., 1995; Gash et al., 1996). For instance dopamine cells are positively influenced by substances like EGF (epidermal growth factor), FGF (fibroblast growth factor) or GDNF (glial cell line derived neurotrophic factor). Gradually the field is moving to applications in human brain diseases. Concerning PD particularly GDNF appears to be a favourable candidate. One of the technical problems to overcome is the route of application. Systemic administration is not possible due to unwanted side effects. Only local continuous application of minute quantities can be the basis of effective and safe therapy. Currently the technique of implanting capsules containing genemodulated cells delivering specific NTF's are at an experimental level but seem promising.

Recently another approach has been tried using osmotic minipumps in non-human primate brain perfusing locally the left striatum with various NTF's during 6 months after inducing parkinsonism using unilaterlal MPTP application (collaboration between the Fondacion Jimenez Diaz, Madrid, Spain and the Paul Scherrer Institute, Villigen, Switzerland). Preliminary results are listed in the Table 1. The animals who received only vehicle fluid or EGF did not improve their FDOPA uptake. Rather the impairment of striatal decarboxylase deteriorated further. On the other hand the animals perfused with FGF in either form did show a clear cut improvement of specific FDOPA uptake even though the lesions in these two animals were severe. The non-perfused striatum also showed a remarkable improvement, resulting in even higher than normal values. These results will be reported in detail elsewhere (de Yebenes et al., in preparation), but serve here to show that local perfusion of NTF may indeed develop into a technique to restore or protect dopamine neurons.

Table 1. Specific FDOPA uptake (Ki) (unit: 1/min) in 6 cynomolgus monkeys

		R Striatum	L Striatum
1. Healthy		0.0138	0.0148
2. Vehicle	before	0.0016	0.004
	after	0.0015	0.002
3. EGF	before	0.007	0.004
	after	0.002	0.001
4. aFGF	before	0.008	0.001
	after	0.025	0.005
5. bFGF	before	0.008	0.001
	after	0.029	0.004

The left side of the monkeys 2 to 5 were perfused with MPTP. FDOPA PET scans were performed before and 6 months after local treament with one NTF in each monkey. Monkey no. 2 received vehicle fluid. See text

Implantation

It has been shown by several groups that dopamine cells from embryo's with a gestation age between 6 to 9 weeks can be successfully implanted into striatum of PD patients leading to partially recovery of dopadecarboxylase capacity in the ensuing months accompanied by various degrees of clinical improvement. Only little post-mortem data is available but these show that the grafted cells had been accepted by the host and axonal outgrowth had taken place directed to projection neurons distributed over large sections of the striatum. The time course of recovery coincided in most studies with local improvement of specific FDOPA uptake measured using PET. If no graft acceptance or clinical improvement had taken place, also no improvement of FDOPA tracer uptake was demonstrated. The to date available implantation data underscores that FDOPA uptake relates to the dopaminergic nerve terminal pool in the striatum. This means that these PET tracer methods play an important complementary role in assessing the various implantation proto-cols which are currently under investigation.

References

Eidelberg D, Moeller JR, Dhawan V, Spetsieris P, Takikawa S, Ishikawa T, Chaly T, Robeson W, Margouleff D, Przedborski S, Fahn S (1994) The metabolic anatomy of Parkinsonism. J Cereb Blood Flow Metabol 14: 783–801

Eidelberg D, Moeller JR, Ishikawa T, Dhawan V, Spetsieris P, Chaly T, Belakhlef A, Mandel F, Przedborski S, Fahn S (1995a) Early differential diagnosis of Parkinson's disease with 18F-fluorodeoxy-glucose and positron emission tomography. Neurology 45: 1995–2004

Eidelberg D, Moeller JR, Ishikawa T, Dhawan V, Spetsieris P, Chaly T, Robeson W, Dahl R, Margouleff D (1995b) Assessment of disease severity in Parkinsonism with fluorine-18-fluorodeoxy-glucose and PET. J Nucl Med 36: 378–383

de Rijk MC, Breteler MMB, Graveland GA, Ott A, van der Meché, Hofman A (1995) Prevalence of Parkinson's disease in the elderly: the Rotterdam study. Neurology 45: 2143–2146

Gash DM, Zhang Z, Ovadia A, Cass WA, Yi A, Simmerman L, Russell D, Martin D, Lapchak PA, Collins F, Hoffer BJ, Gerhardt GA (1996) Functional recovery in parkinsonian monkeys treated with GDNF. Nature 380: 252–255

Leenders KL (1990) Cerebral energy metabolism and blood flow in Parkinson's disease. In: Martin WRW (ed) Functional imaging in movement disorders. CRC Press, Boca Raton Ann Arbor, Boston, pp 115–130

Leenders KL, Palmer AJ, Quinn N, Clark JC, Firnau G, Garnett ES, Marsden CD (1986) Brain dopamine metabolism in patients with Parkinson's disease measured with positron emission tomography. J Neurol Neurosurg Psychiat 49: 853–860

Leenders KL, Salmon EP, Turton D, Tyrrell P, Perani D, Brooks DJ, Sagar H, Jones T, Marsden CD, Frackowiak RSJ (1990) The nigrostriatal dopaminergic system assessed in vivo by positron emission tomography in healthy volunteer subjects and patients with Parkinson's disease. Arch Neurology 47: 1290–1298

Marie RM, Rioux P, Eustache F, Travere JM, Lechevalier B, Baron JC (1995) Clues about the functional neuroanatomy of verbal working memory: a study of resting brain glucose metabolism in Parkinson's disease. European J Neurol 2: 83–94

Morrish PK, Sawle GV, Brooks DJ (1995) Clinical and [F-18]dopa PET findings in early Parkinson's disease. J Neurol Neurosurg Psychiat 59: 597–600

Patlak CS, Blasberg RG (1985) Graphical evaluation of blood-to-brain transfer constants from multiple-time uptake data. Generalizations. J Cereb Blood Flow Metabol 5: 584–590

Tomac A, Lindqvist E, Lin LFH, Ögren SO, Young D, Hoffer BJ, Olson L (1995) Protection and repair of the nigrostriatal dopaminergic system by GDNF in vivo. Nature 373: 335–341

Author's address: Prof. Dr. K. L. Leenders, Paul Scherrer Institute, CH-5232 Villigen, Switzerland.

Contributions of Positron Emission Tomography to elucidating the pathogenesis of Idiopathic Parkinsonism and Dopa Responsive Dystonia

D. B. Calne, R. de la Fuente-Fernández, and **A. Kishore**

Neurodegenerative Disorders Centre, Division of Neurology, Department of Medicine, The University of British Columbia, Vancouver Hospital and Health Sciences Centre, Vancouver, Canada

Summary. The metabolic mapping of brain activity, using PET, confirms the conventional wisdom of neurophysiology. In studies of pathophysiology, PET has yielded evidence that has generated new hypotheses. Progression of the lesion detectable with fluorodopa, in human subjects exposed to MPTP, raises the possibility of a transient environmental event being a cause of progressive neurodegeneration. Studies with fluorodopa in Idiopathic Parkinsonism indicate that the rate of loss of neurons is faster initially, and then tends to approach the normal age-related decline. In dopa responsive dystonia, the finding of normal fluorodopa PET led to the prediction that the lesion would be functional rather than anatomical; this has been confirmed by the identification of a defect in dopamine synthesis in this disorder. Filally, new PET ligands show promise for future studies designed to unravel the pathogenesis of diseases involving the basal ganglia.

Introduction

Positron Emission Tomography has yielded evidence of considerable importance to movement disorders. Studies with O^{15}, and with F^{18} fluorodeoxyglucose, have demonstrated increased metabolism in the lenticular nucleus contralateral to Parkinsonian deficits (Martin et al., 1984; Wolfson et al., 1985). In the light of current understanding of the physiology of the basal ganglia, the increased lenticular activity may well stem from augmentation in pallidal metabolism. This same increase in pallidal activity provides the rationale for placing lesions in the pallidum of patients with Parkinsonism. In contrast to this hyperactivity, there is normally hypometabolism in the cerebral cortex of patients with Parkinsonism. This reduced metabolism is mild in patients with normal intellectual function, but in Parkinsonian subjects who have dementia, the reduced cerebral metabolism is comparable in distribution and severity to the changes found in Alzheimer's disease (Eidelberg et al., 1990; Otsuka et al., 1991; Peppard et al., 1990, 1993).

Thus the metabolic map in Parkinsonism, derived from PET studies, confirms the conventional wisdom. The pathophysiology of the motor deficits derives from increased impulse traffic in the pallidum, and the intellectual impairment being associated with reduced cerebral cortical function. Observations on functional imaging also confirm the role of regions such as the supplementary motor area (Brooks, 1993). In contrast with the role of PET in confirming hypotheses of pathophysiology, its application to explore pathogenesis has yielded evidence from which new hypotheses have been developed. A particularly useful comparison has been made between two movement disorders associated with reduced dopaminergic transmission in the striatum – Idiopathic Parkinsonism (IP) and Dopa Responsive Dystonia (Segawa's disease).

PET ligands for studying pathogenesis

Most work on the pathogenesis of movement disorders has been undertaken with fluorodopa. The value of this ligand has been validated by studies on monkeys, and on human subjects. In monkeys, it has been possible to establish a series of animals with experimental nigrostriatal dopaminergic lesions of differing severity. This population of animals has been created by injecting various doses of MPTP into the carotid artery on one side. Mathematical modelling has allowed quantitative indices to be developed from the PET images of animals who have experimental damage to the nigrostriatal system. Subsequent quantitative histological studies on the brains of these monkeys have allowed correlations to be sought between the quantitative index of fluorodopa PET, and the nigral cell count of dopaminergic neurons. A direct linear relationship exists, indicating that fluorodopa provides an excellent measure of the structural integrity of the nigrostriatal system (Pate et al., 1993).

A similar approach has been adopted to seek correlation between quantitative PET and nigral neuronal counts in human subjects. A series of patients with various disorders involving the nigrostriatal pathway have undergone PET scans, and when these subjects eventually die, their brains have been examined post mortem, with counts of nigral cells of dopaminergic neurons. In spite of the quite variable time interval between PET scanning and death, a direct linear correlation has been found, indicating that in human subjects, quantitative PET indices provide a measure of the integrity of the dopaminergic nigrostriatal system (Snow et al., 1993a).

These studies provide a "proof of principle" that in both animals and human subjects, fluorodopa PET can provide an in vivo index of the structural integrity of the nigrostriatal system.

In recent years, another approach has been introduced to apply PET to the study of pathogenesis in movement disorders. C^{11} raclopride is a ligand that binds to the D2 receptors, and mathematical modelling allows a semi-quantitative measure of uptake (Sawle et al., 1993; Snow et al., 1993b).

Application of PET to study the pathogenesis of Idiopathic Parkinsonism

With this background, it has been possible to explore the relationship between various clinical measures of Parkinsonism, and the integrity of the nigrostriatal system as measured by PET. A direct linear relationship has been found for the bradykinesia score of the Modified Columbia and UPDRS protocols (Eidelberg et al., 1990; Lee et al., 1994). A similar relationship has been found for measures of motor function made with the Purdue pegboard test. Correlations exist between PET measures and rigidity, but these are not as consistent as those for bradykinesia. The relationship between PET measurements and tremor scores is even less consistent.

From these findings, it becomes evident that we have a unique clinical measure for IP, which is linearly related to the severity of the underlying pathology. Thus a reduction on the bradykinesia scores by fifty percent correlates with a decrease in the PET measures with fluorodopa, also by fifty percent, and this in turn indicates that there has been a fifty percent loss of dopaminergic nigrostriatal neurons. From this background, it is possible to survey large numbers of patients and make measurements from which we can infer the quantitative status of the nigrostriatal system. With this analytic system, we have studied patients with different levels of severity and duration of IP. We have found that progression is more rapid initially, and eventually, after about fifteen years of symptoms, the rate of advance of the illness seems to derive entirely from the normal age related attrition of nigrostriatal cells. From the curve of the natural history of neuronal loss, obtained in this way, we can also extrapolate backwards an estimate that IP generally begins some five years before symptoms are first noticed. The decelerating rate of progression of dopaminergic nigral cell loss in IP is important since any hypothesis of pathogenesis must account for this natural history. On the basis of this finding, and other related observations, it has been suggested that IP might be initiated by a transient etiological event superimposed on a genetic predisposition (Lee et al., 1994; Schulzer et al., 1994). The event could kill some dopaminergic cells, and damage others so that their life expectation was reduced. This notion contrasts with conventional concepts of the pathogenesis of IP continuously engaging healthy neurons.

In addition to the theoretical importance of these new ideas, they have a practical significance in guiding us towards the type of etiology that we should be seeking. There is also an important implication so far as treatment is concerned. Pharmacotherapy for the pathogenesis should be directed at rescuing damaged cells rather than protecting healthy cells. Furthermore, surgical approaches to treatment, such as transplantation, would potentially be curable if the etiology of IP derives from a transient event. In contrast, if there is a continuous destructive process engaging healthy neurons, transplantation of dopaminergic nerve cells would simply be palliative, since in due course the implanted neurons would also be attacked by the pathogenic process.

The application of PET with C^{11} raclopride has yielded interesting results in Idiopathic Parkinsonism. In the early stages of pathogenesis, loss of nigrostriatal dopaminergic nerve fibres leads to upregulation of the postsynaptic D2 receptors that bind to raclopride in the putamen. Thus as fluorodopa binding falls, raclopride binding increases (Rinne et al., 1990; Sawle et al., 1993; Snow et al., 1993b). This allows the ratio between fluorodopa uptake and raclopride uptake to be employed as the most sensitive index of degeneration in the nigrostriatal system. In more advanced Idiopathic Parkinsonism, the upregulation of raclopride binding is no longer evident (Brooks et al., 1992; Shinotoh et al., 1993). This may reflect the extension of pathology into neuronal components postsynaptic to the nigrostriatal projection, or alternatively, the use of dopaminomimetic drugs as treatment for advanced Idiopathic Parkinsonism may lead to down-regulation of dopamine receptor binding.

The application of PET to the study of the pathogenesis of Dopa Responsive Dystonia

PET has played a central role in unravelling of the pathogenesis of Dopa Responsive Dystonia. A crucial problem, in early studies of Dopa Responsive Dystonia, was the considerable difficulty in distinguishing this disorder from Juvenile Parkinsonism. In both situations, there is a generally a combination of Parkinsonism and Dystonia. Both disorders respond to levodopa. Early PET studies, with fluorodopa, led to conflicting results, and in retrospect it is virtually certain that this confusion was caused by incorrect clinical diagnoses. All investigators who applied PET to the study of Dopa Responsive Dystonia encountered the same problem. At a small symposium held by Dr. Segawa in 1990, stringent criteria were developed for the clinical diagnosis of Dopa Responsive Dystonia (Calne et al., 1993). and these criteria allowed the disorder to be analyzed by PET. A series of careful studies by Snow et al. (1993c) demonstrated quite clearly that in Dopa Responsive Dystonia fluorodopa uptake is normal, in contrast to the reduced striatal uptake of fluorodopa that occurs in Juvenile Parkinsonism. So in Dopa Responsive Dystonia the salient observations are that, 1) the disorder responds to levodopa; 2) fluorodopa uptake is normal.

In the light of studies showing that fluorodopa uptake reflects the integrity of the nigrostriatal system, it was proposed that the pathogenesis of Dopa Responsive Dystonia was associated with a functional deficit in the nigrostriatal system, leading to reduced dopaminergic transmission *without* loss of dopaminergic nigrostriatal neurons. This conclusion was supported by a post mortem study of Dopa Responsive Dystonia reported by Rajput et al. (1994). With this background, it became clear that Dopa Responsive Dystonia was associated with a functional disturbance of the nigrostriatal system rather than a structural change. In 1994, Nagatsu and his colleagues (Ichinose et al., 1994) confirmed this hypothesis by demonstrating a defect in the gene for GTP cyclohydrolase 1, the enzyme responsible for the synthesis of tetrahydrobiopterin. Tetrahydrobiopterin is the co-factor for tyrosine hydroxylase, so the hereditary deficit in Dopa Responsive Dystonia impedes synthesis of dopamine through a deficit in hydroxylation of tyrosine, which is the rate limiting step in the synthetic pathway. Two out of the three initial families studied with Dopa Responsive Dystonia had the genetic abnormality described by Nagatsu and his colleagues. Recently, a distinct but functionally similar mutation has been found to account for other cases of Dopa Responsive Dystonia (Mizuno, personal communication).

Thus PET provided a precise prediction of the nature of the pathogenesis of Dopa Responsive Dystonia and, in particular, it clearly indicated that the nature of the pathogenesis was quite distinct from that of Juvenile Parkinsonism. Incidentally, PET studies and post mortem investigations have all indicated that Juvenile Parkinsonism resembles adult Idiopathic Parkinsonism so far as pathogenesis is concerned.

There remained one area for exploration in Dopa Responsive Dystonia. What happens to the D2 receptor sites in this condition? Studies with C^{11} raclopride have provided an answer. It has been rather difficult to perform PET studies here, because as soon as the diagnosis of Dopa Responsive Dystonia is made, patients are treated with dopaminomimetic agents, which are known to modify D2 receptor binding. In these circumstances, studies have to be confined to the occasional untreated case of Dopa Responsive Dystonia, or to the investigation of obligate carriers of the gene for the disorder. In both categories of subject, the binding of C^{11} raclopride is elevated, indicating that there is upregulation of D2 receptors in Dopa Responsive Dystonia (Kishore et al., 1995).

Conclusions

Taking Idiopathic Parkinsonism and Dopa Responsive Dystonia together, we find a coherent picture of the distinct and different pathogenesis of these disorders revealed by PET studies. In the future, we can look forward to more extensive exploration of the disturbances in the neurobiology of dopaminergic transmission in pathology attacking the nigrostriatal pathway. This extension in our knowledge will be achieved through the use of new ligands that will allow further analysis of the various components of the dopaminergic synapse. For example, PET is now possible with an analog of tetrabenazine, that binds to dopamine vesicular storage membranes, and several ligands are currently being studied for the dopamine transporter of the nigrostriatal nerve endings. These include derivatives of β-CIT, and of methylphenidate. PET studies with these new ligands will be correlated with post mortem findings to provide new insight concerning the ways in which the nigrostriatal pathway can be damaged by disease, and perhaps the ability to modify the underlying pathogenesis by novel approaches to treatment.

References

Brooks DJ (1993) Functional imaging in relation to parkinsonian syndrome. J Neurological Sci 115: 1–17

Brooks DJ, Ibanez V, Sawle G, et al (1992) Striatal D2 receptor status in patients with Parkinson's disease striatonigral degeneration and progessive supranuclear palsy, measured with 11 C-Raclopride and positron emission tomography. Ann Neurol 31: 184–192

Calne DB, Nygaard TG, Snow BJ (1993) The distinction between early onset idiopathic parkinsonism (Juvenile Parkinson's disease) and dopa-responsive-dystonia (hereditary progressive dystonia, Segawa dystonia). In: Segawa M (ed) Hereditary progressive dystonia with marked diurnal fluctuation. The Parthenon Publishing Group, Carnforth, UK, pp 215–218

Eidelberg D, Moeller JR, Dhawan V, et al (1990) The metabolic anatomy of Parkinson's disease: Complementary [18F] fluorodeoxyglucose and [18F] fluorodopa positron emission tomographic studies. Mov Disord 5: 203–213

Ichinose H, Ohye T, Takahashi E, et al (1994) Hereditary progressive dystonia with marked diurnal fluctuation caused by mutations in the GPT cyclohydrolase I gene. Nature Genetics 8: 236–242

Kishore A, Snow BJ, Naini AB, Przedborski S, Vingerhoets FJG, Nygaard TG (1995) Analysis of dopaminergic function in asymptomatic carriers of the DRD gene: CSF dopaminergic metabolites and PET with fluorodopa and raclopride. Neurology 45: A187 (Abstract)

Lee CS, Schulzer M, Mak EK, et al (1994) Clinical observations on the rate of progression of idiopathic parkinsonism. Brain 117: 501–507

Martin WRW, Beckman JH, Calne DB, et al (1984) Cerebral glucose metabolism in Parkinson's disease. Can J Neurol Sci 11 [Suppl 1]: 169–173

Otsuka M, Ichiya Y, Hosokawa S, et al (1991) Striatal blood flow, glucose metabolism and 18F-dopa uptake: Difference in Parkinson's disease and atypical Parkinsonism. J Neurology, Neurosurgery, and Psychiatry 54: 898–904

Pate BD, Kawamata T, Yamada T, et al (1993) Correlation of striatal fluorodopa uptake in MPTP monkey with dopaminergic indices. Ann Neurol 34: 331–338

Peppard RF, Martin WRW, Clark CM, Carr GD, McGeer PL, Calne DB (1990) Cortical glucose metabolism in Parkinson's and Alzheimer's disease. J Neurosci Res 27: 561–568

Peppard RF, Martin WRW, Carr GD, et al (1993) Cerebral glucose metabolism in Parkinson's disease with and without dementia. Arch Neurol 49: 1262–1268

Rajput AH, Gibb WRG, Zhong XH, et al (1994) Dopa-responsive dystonia: Pathological and biochemical observations in a case. Ann Neurol 35: 396–402

Rinne UK, Laihinen A, Rinne JO, Nagren K, Bergman J, Ruotsalainen U (1990) Positron emission tomography demonstrates dopamine D2 receptor supersensitivity in the striatum of patients with early Parkinson's disease. Mov Disord 5: 55–59

Sawle GV, Playford ED, Brooks DJ, Quinn N, Frackowiak SJ (1993) Asymmetrical pre-synaptic and post-synaptic changes in the striatal dopamine projection in dopa naive parkinsonism. Brain 116: 853–867

Schulzer M, Lee CS, Mak EK, Vingerhoets FJG, Calne DB (1994) A mathematical model of pathogenesis in idiopathic parkinsonism. Brain 117: 509–516

Shinotoh H, Hirayama K, Tateno Y (1993) Dopamine D1 and D2 receptors in Parkinson's disease and striatonigral degeneration determined by PET. In: Narabayashi H, Nagatsu T, Yanagisawa N, Mizuno Y (eds) Parkinson's disease from basic research to treatment, Advances in Neurology, vol 60. Raven Press, New York, pp 488–493

Snow BJ, Tooyama I, McGeer EG, et al (1993a) Human positron emission tomographic [18F] fluorodopa studies correlate with dopamine cell counts and levels. Ann Neurol 34: 324–330

Snow B, Buckley K, Bailey D, et al (1993b) Dopamine-D2 receptor density is inversely correlated with dopaminergic innervation in untreated Parkinson's disease. Neurology 43: A269

Snow BJ, Nygaard TG, Takahashi H, Calne DB (1993c) Positron emission tomographic studies of dopa-responsive dystonia and early-onset idiopathic parkinsonism. Ann Neurol 34: 733–738

Wolfson LI, Leenders KL, Brown LL, Jones T (1985) Alterations of regional cerebral blood flow and oxygen metabolism in Parkinson's disease. Neurology 35: 1399

Authors' address: Prof. Dr. D. B. Calne, Neurodegenerative Disorders Centre, Vancouver Hospital and Health Sciences Centre, Purdy Pavillon, Room M 36, 2221 Wesbrook Mall, Vancouver, BC, Canada V6T 2B5.

Endogenous and exogenous neurotoxins

Chair: M. B. H. Youdim and P. Riederer

Mechanism of 6-hydroxydopamine neurotoxicity

Yelena Glinka, M. Gassen, and **M. B. H. Youdim**

Department of Pharmacology, Bruce Rappaport Family Research Institute,
Faculty of Medicine, Technion, Haifa, Israel

Summary. The catecholaminergic neurotoxin 6-hydroxydopamine (6-OHDA) has recently been found to be formed endogenously in patients suffering from Parkinson's disease. In this article, we highlight the latest findings on the biochemical mechanism of 6-OHDA toxicity. 6-OHDA has two ways of action: it easly forms free radicals and it is a potent inhibitor of the mitochondrial respiratory chain complexes I and IV. The inhibition of respiratory enzymes by 6-OHDA is reversible and insensitive towards radical scavengers and iron chelators with the exception of desferrioxamine. We conclude that free radicals are not involved in the interaction between 6-OHDA and the respiratory chain and that the two mechanisms are biochemically independent, although they may act synergistically in vivo.

1. Introduction

6-Hydroxydopamine (6-OHDA) is a catecholaminergic neurotoxin, which is employed for the experimental study of Parkinson's disease in animal models. For a long time it was regarded as an exogenous toxin only, and thus the biochemical mechanism of its toxicity was of limited interest. Recently however, Andrew et al. (1993) found endogenous 6-OHDA in the urine of parkinsonian patients. This finding supports a possible involvement of this catecholamine in the pathogenesis of Parkinson's disease and confirms the high relevance of biochemical studies of 6-OHDA toxicity.

2. Potential pathogenic mechanisms: Oxidative stress and impairment of energy metabolism

2.1 Free radicals in neurodegenenation

Parkinson's disease is characterized by the progressive degeneration of the dopaminergic neurons originating in the substantia nigra (pars compacta). Although the reason of this selective cell death has not been understood, several mechanistic concepts have been put forward. Since typical indicators of oxidative stress (increased basal lipid peroxidation, decrease in reduced glutathione, increase of iron content) were determined in postmortem brains of parkinsonian patients, the pathogenesis of Parkinson's disease was thought to be based mainly on the damage of dopaminergic neurons by oxygen derived free radicals (Youdim and Riederer, 1993). Similarly the toxicity of 6-OHDA is proposed to be related to its ability to produce free radicals and to cause oxidative stress (Cohen and Werner, 1994). The damage induced by 6-OHDA in cultured neuronal cells is similar to the effects of H_2O_2: it affects multiple targets including the uptake sites for glucose and α-aminoisobutyric acid (Vroegop et al., 1995). The availability of reduced glutathione is decreased after 6-OHDA intoxication (Perumal et al., 1992), and the activation of glutathione reductase in presence of exogenous brain-derived neurotrophic factor promoted the survival of cultured dopaminergic neurons (Spina et al., 1992; Altar et al., 1993). In vivo, this trophic support is mainly provided by glia cells. Stimulated by 6-OHDA, glia were found to release basic fibrillary growth factor (Chadi et al., 1994), nerve growth factor, brain derived neurotrophic factor and proenkephalin (Schwartz and Nishiyama, 1994) as well as glia derived neurotrophic factor (Offen et al., 1995). Neurotrophins do not only improve the availability of reduced glutathione, they also raise catalase and glutathione peroxidase levels (Sampath et al., 1994) and thus increase the cellular defense against oxidative stress. The induction of neurotrophins by 6-OHDA can be free radical mediated, as treatment of cultured astrocytes with H_2O_2 brings about the same effect (Naveilhan et al., 1994).

Iron is an important factor in the process of free radical formation and propagation. The co-localization of lesions and iron-enriched zones in postmortem brains (Dexter et al., 1992), melanin involvement in iron accumulation by dopaminergic neurons (Ben-Shachar and Youdim, 1990) and induction of animal parkinsonism by stereotactic intrabrain iron injections (Ben-Shachar and Youdim, 1991) indicates iron involvement in the pathogenesis of Parkinson's disease. Redox active metal ions like Fe^{2+}/Fe^{3+} or Cu^+/Cu^{2+} can enhance oxidative stress through induction of hydroxyl radical ($HO^\bullet$) formation from H_2O_2 via Fenton-type reactions. Iron dependent mechanisms and free radicals may also contribute to the toxicity of 6-OHDA, which has been shown to liberate iron from ferritin (Monterio and Winterbourne, 1989) and to increase the availability of ferrous iron (Fe^{2+}) for the Fenton reaction. Such a mechanism could explain the finding that the iron chelator desferrioxamine provides protection against brain lesioning induced by 6-OHDA injections in rats (Ben-Shachar et al., 1991).

2.2 Biochemical consequences of insufficient energy supply in neural cells

An increasing body of evidence points out the correlation between neurodegenerative diseases like Parkinson's disease or Alzheimer's disease and reduced activity of the mitochondrial and cytosolic enzymes involved in energy metabolism (for review see Beal et al., 1993; Mizuno et al., 1993). Although the sequence of toxic events starting from impairment of mitochondrial function is still far from being understood, several sensitive steps in cell metabolism can be pointed out. Respiratory failure affects the plasmatic membrane potential: cell depolarization releases the Mg^{2+} block of the voltage sensitive NMDA receptors. This results in increased Ca^{2+} influx stimulated by the excitotoxic amino acid neurotransmitters glutamate and aspartate. With the Mg^{2+}-block released, extracellular glutamate is able to stimulate uncontrolled calcium entrance through the gate and bring about neural degeneration. This mechanism is discussed in the excellent review of Beal et al. (1993).

Inhibition of respiration obviously affects all energy dependent processes. In nerve cells this is especially neurotransmission, which requires significant amounts of energy for neurotransmitter uptake by vesicles, for storage and transport, transmitter re-uptake by the presynaptic neuron, and axonal transport in both directions. The required ATP is mostly provided by mitochondrial respiration (Erecinska and Dagani, 1990). Hence, impaired respiration and ATP depletion can cause cytosolic and extracellular accumulation of neurotransmitters and breakedown of axonal transport. Catecholamines themselves are potent inhibitors of mitochondrial respiration (Ben-Shachar et al., 1994) and their non-compartmentalized accumulation can augment the cellular energy deficit. Furthermore, their MAO-catalyzed metabolism as well as their non-enzymatic oxidation contributes to free radical production. Dopamine toxicity for both pre- and postsynaptic neurons was pointed out by Filloux and Townsend (1993).

The finding that mitochondrial respiration is impaired in parkinsonian patients (Mizuno et al., 1989) gives rise to the question whether reactive oxygen species are primarily responsible for neurodegeneration. Morphological and biochemical studies of postmortem brains give a clear account of oxidative tissue damage (Youdim and Riederer, 1993). However, it still remains unclear, whether free radicals causing damage to membrane lipids and proteins are the underlying factor to iniciate the failure of mitochondrial respiration, deterioration of membrane potentials and disturbance of calcium homeostasis, or whether oxidative stress occurs just as an accompanying phenomenon of neurodegeneration. A sequence of events starting with an inhibition of mitochondrial respiration may deplete cellular energy sources with a decrease of membrane potentials leading to the disturbance of calcium homeostasis (Beal et al., 1993) and insufficient cellular protection against free radical damage due to the lack of NADPH for glutathione reduction. This will give rise to the activation of calcium-dependent proteases and lipases and increased formation of free radicals and bring about cell death (Orrenius and Nicotera, 1994). The final results in both cases will be cell death linked with biochemical evidence for oxidative stress. A way to distinguish between these two mechanisms will be the use of pharmacological tools that selectively inhibit critical steps. Meanwhile for effective therapy and preventive protection strategy the knowledge of the true consequence of the fatal processes is of great value.

In this article, we will discuss the mechanism of 6-OHDA neurotoxicity. A key question will be, whether the its inhibition of respiratory enzymes (Glinka and Youdim, 1995) is a free radical dependent process and whether the 6-OHDA induced damage to dopaminergic neurons is mediated by free radicals.

3. 6-OHDA: What is it doing in the cell?

3.1 What is the role of free radicals in 6-OHDA toxicity?

The most prominent biochemical property of 6-OHDA is its inhibition of brain mitochondrial complexes I and IV (Glinka and Youdim, 1995). Our experiments to elucidate the mechanism of inhibition were especially aimed to confirm or to rule out the involvement of free radicals. 6-OHDA itself is easily oxidizable to form semiquinone radicals and can participate in the generation of reactive oxygen species. If this is in any way related to its toxicity and the loss of the respiratory activity results from oxidative modification of the enzymes by free radicals, any conditions facilitating oxidation and radical formation could enhance the toxic effect of this catecholamine. Especially iron is expected to increase the effect by the activation of the formation of free radicals. The inhibition should be decreased by antioxidants and by iron chelators. Our experiments (Glinka et al., 1996b), however clearly showed that a) the inhibition is not prevented by any anti-oxidants tested; b) elevated iron concentrations do not increase the inhibition but decrease it, most likely by interaction with the inhibitor 6-OHDA c) decrease of the iron level in vivo in iron-deficient animals or in vitro by iron chelation does not protect mitochondrial complex I against inhibition by this toxin (the only exception, desferrioxamine, will be discussed later in the text) and most importantly: d) the inhibition is fully reversible.

On the basis of this data it can be completely ruled out that the inhibition of the respiratory complex I and IV by 6-OHDA depends on free radicals. Still, it cannot be ruled out that impaired mitochondrial respiration leads to an increase of free radical production. Many proteins are relatively insensitive to oxidant attack. However if protein oxidation occurs, it is difficult to prevent it by antioxidants, as it occurs in very close proximity to metal binding sites (Stadtman, 1993). Any increase of oxidative stress by 6-OHDA however does not disactivate complex I and IV, as this effect would not be reversible. We concluded, that the mitochondrial toxicity of 6-OHDA must be mediated by non-radical interaction of 6-OHDA with this enzyme, and that the inhibition of mitochondrial respiration by this toxin brings about neurodegeneration via ATP depletion. This interpretation is further supported by the in vivo finding that rat mortality induced by intrabrain dopamine injections did not correlate with the state of oxidative stress in animals brains. Dopamine is able to induce apoptosis in primary thymocyte cell culture (Offen et al., 1995), which may be due to its inhibition of mitochondrial complex I. This inhibition was regarded as a likely source of fatal brain damage (Ben-Shachar et al., 1994). Thus, inhibition of mitochondrial respiration may be regarded as a primary event, which then starts the sequence of toxic processes resulting in cell death.

Although iron does not enhance 6-OHDA toxicity in mitochondrial respiration, other divalent cations increase its inhibitory potency. 0.5 μM Ca^{2+} decreased the IC_{50} for 6-OHDA by 25% and 100 μM Ca^{2+} by 40%. Zinc shows a similar effect: 20 μM cause an approximately 40% increase of the enzyme's sensitivity to inhibition, while 20 μM Fe^{3+} render it copletely insensitive (see

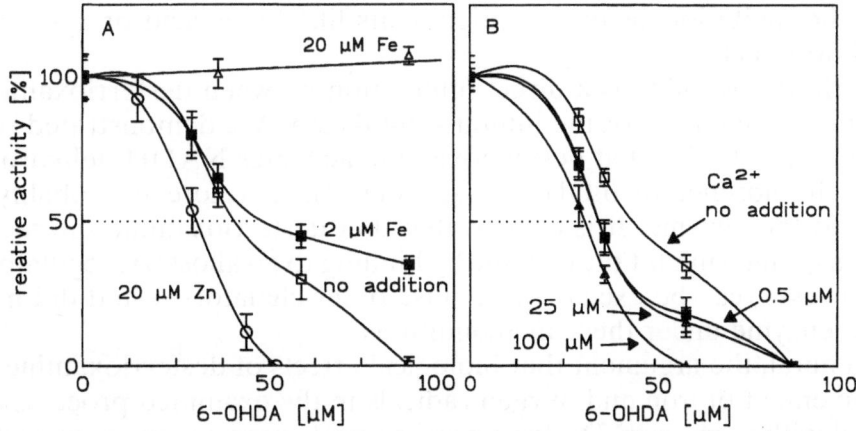

Fig. 1. The effect of metal ions (A: Fe^{3+}, Zn^{2+}; B: Ca^{2+}) on the sensitivity of brain mitochondrial NADH-coenzyme Q reductase towards inhibition by 6-OHDA

Fig. 1). These cations inhibit the slow transition of the complex I molecule to its active conformation (Kotlyar et al., 1992), thus increasing the percentage of the inactive enzyme. This process probably interacts with the inhibition caused by 6-OHDA. Several cations including manganese, nickel, and barium slow down the enzyme activation. Iron is likely to show a similar effect, but no data is available. A slight potentiation of the inhibition, which might be expected for iron, might be masked completely by iron chelation or oxidation of the inhibitor.

The calcium-dependent augmentation of sensitivity of NADH dehydrogenase (complex I) to the inhibition by 6-OHDA may cause a feed-back amplification under certain conditions, when the inhibition of the mitochondrial respiration enhances calcium entrance into the cell and the elevation of calcium concentration further increases the inhibition of mitochondrial complex I. This vicious circle can be interrupted by calcium chelators or blockers of NMDA receptors, but this will not prevent the inhibition of complex I, which appears to initiate the chain of the toxic events.

3.2 Protection of mitochondrial complex I by desferrioxamine

Regarding the convincing evidence in favor of a non-radical mechanism of 6-OHDA toxicity, the finding that the iron chelator desferrioxamine shows a remarkable protective effect in vitro as well as in vivo deserves a more thorough discussion. In contrast to other chelators like EDTA, 2,2-dipyridil and D-penicillamine, desferrioxamine prevented inhibition of NADH dehydrogenase by 6-OHDA and protected rats from death induced by intrabrain injections of 6-OHDA in vivo (Ben-Shachar et al., 1991). It appears, that properties of desferrioxamine other than metal chelation are of importance here. Its antioxidant effect is well documented (Morel et al., 1992) but most likely does not play any role in the protection of NADH dehydrogenase against the inhibition, as other antioxidants like lipoic acid or apomorphine showed no effect.

The third possibility is a direct interaction between desferrioxamine and either the enzyme itself or the inhibitor 6-OHDA. We demonstrated recently (Glinka et al., 1996b), that desferrioxamine activates NADH dehydrogenase even in the absence of 6-OHDA (see Fig. 2), therefore it probably binds directly to the enzyme. It is conceivable that desferrioxamine interacts with metal ions, which inhibit the enzyme by binding to an allosteric regulatory site. This, however, can be excluded, because other chelators tested did not activate the enzyme under the same conditions.

In general, the argument that biological effects of desferrioxamine reflect an involvement of iron and oxygen radicals in the examined processes has to regarded with more care. Desferrioxamine protection against paraquat toxicity is based in part on its ability to prevent the uptake of the toxin by alveolar cells via the polyamine site, and the same effect can be observed with putrescine taken in the same concentrations (Ben-Shachar et al., 1991). This is another example of a desferrioxamine effect, which is not related to iron chelation.

Addition of equimolar calcium or iron completely prevented the protection of complex I by desferrioxamine, though the same concentrations of iron-EDTA complex did not show any effect (see Fig. 3). It appears, that the cyclic conformation of the desferrioxamine-metal chelate is probably unable to bind to the enzyme, and that in contrast the open linear conformation can bind and protect. Further investigations demonstrated a tenfold increase of the inhibitor constant for 6-OHDA in the presence of desferrioxamine. The kinetic analysis revealed that there is no competition between inhibitor and protector, but rather an allosteric interaction: the binding of the protector makes the inhibitor binding unfavorable (Glinka et al., 1996a).

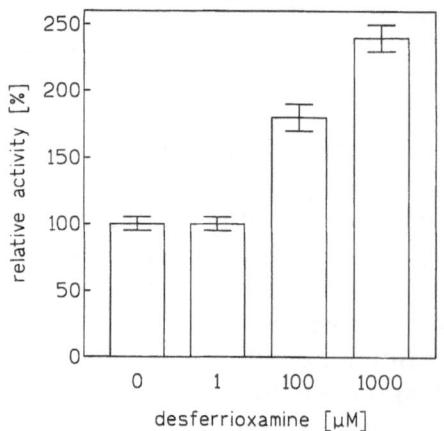

Fig. 2. Activation of brain mitochondrial NADH-coenzyme Q reductase by desferrioxamine (measured as NADH turnover in the presence of decylubiquinone, a water soluble ubiquinone analog, see Glinka et al., 1996b)

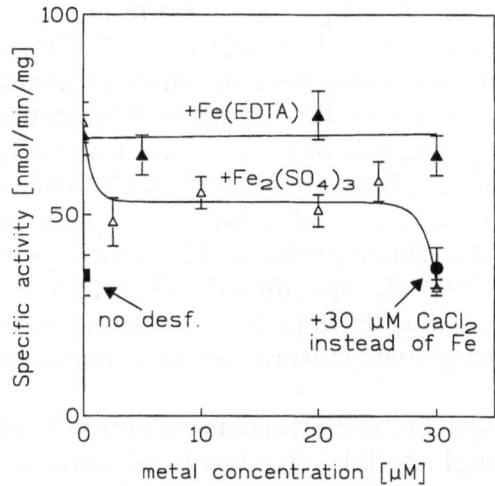

Fig. 3. Desferrioxamine (30 mM) protects brain mitochondrial NADH-coenzyme Q reductase from inhibition by 0.46 mM 6-OHDA. This protection is prevented by free, but not by chelated metal ions

This in vitro protection of complex I against its inhibition by 6-hydroxy-dopamine and the impressive prevention of parkinsonian symptoms in animals after intrabrain injections of 6-OHDA (Ben-Shachar et al., 1991) might open a new way for the development of treatments or preventive strategies for neurodegenerative diseases.

3.3 What are the reasons for the selectivity of 6-OHDA toxicity?

6-OHDA non-selectively inhibits the mitochondrial complex I from various cell types including neural cells and hepatocytes (Glinka et al., 1996b). Consequently, there must be other mechanisms in operation to account for the well known selectivity in causing degeneration of specified brain areas and types of neurons. Since 6-OHDA was shown to compete with dopamine for its uptake site (Decker et al., 1993) and 6-OHDA toxicity was prevented by the dopamine uptake blocker (Cerruti et al., 1993), it seems to depend on the availability of a monoamine carrier for entrance into the presynaptic neurons. This can also account for the selective damage of norepinephrine neurons. The irreversible modification of the dopamine uptake site by 6-OHDA (Decker et al., 1993) obviously results from covalent modification of the protein by either oxygen- or 6-OHDA-derived radicals and may be of increased relevance under conditions facilitating free radical reactions (e.g. elevated iron concentration). This modification completely prevents further dopamine entrance into the cell and thus contributes to the disturbance of neurotransmission (Decker et al., 1993). On the other hand it also prevents the entrance of the toxin and the inhibition of mitochondrial respiration and can thus limit the toxic effects of 6-OHDA.

Although carriers for the uptake of 6-OHDA are available, not all dopaminergic neurons are equally sensitive to 6-OHDA-induced toxicity. Neonatal hypothalamic dopamine neurons show resistance independently of their anatomical and functional differences and the route of injection under conditions that lead to complete degeneration of nigral neurons (Yokoyama et al., 1993). This further confirms that in 6-OHDA toxicity selective pathways are operative, which are inconceivable for a multiple-target non-specific free radical mechanism that has been postulated by Vroegop et al., (1995). According to Gordon et al. (1991), the specific 6-OHDA-induced damage of murine megakariocyte progenitor cells depends on the activation of the catecholamine transporter during their maturation. Less matured cells are less sensitive.

In order to distinguish between radical and non-radical mechanisms of cell damage in experimental models, the levels of antioxidants are of critical importance. Even if free radicals are observed in vitro, the level of antioxidants in vivo available might be sufficient to prevent rapid oxidation of the toxin thus supporting an alternative route of its toxicity.

3.4 Can in-vivo-formed 6-OHDA be relevant for neurodegeneration?

A compound as reactive and oxidizable as 6-OHDA is expected to show its effects only in close vicinity to the site of its formation. In vivo this may be wherever dopamine is present, as 6-OHDA can be formed from dopamine by non-enzymatic hydroxylation in presence of Fe^{2+} and H_2O_2 (Kienzl et al., 1996). 6-OHDA may interact with the variety of the catecholamine binding and uptake sites (Cerruti et al., 1993), and inhibit mitochondrial complex I (Glinka and Youdim, 1995). 6-OHDA is also able to reduce ferritin-bound Fe^{3+} leading to the release of Fe^{2+} (Kienzl et al., 1996). Of course it can also undergo rapid autoxidation with radical formation. 6-OHDA is a substrate for monoamine oxidase (Karoum et al., 1993) and may also compete with dopamine for the vesicular uptake, dopamine β-hydroxylase and catechol-O-methyl transferase (COMT) reactions and can participate in neuromelanin formation. Significant amounts of cytosolic dopamine make non-enzymatic 6-OHDA formation more likely, especially if dopamine metabolism is inhibited by MAO and COMT (Kita et al., 1995). If located in the cytosol, 6-OHDA can reach most cellular targets, including mitochondria. 6-OHDA is incorporated into vesicles and competes with dopamine for the Mg-ATP-dependent vesicular uptake. One can speculate, whether, like for the re-uptake transporter (Cerruti et al., 1993), irreversible modification of the carrier occurs. These effects may decrease the vesicular storage capacity for dopamine and disturb the synaptic function. An increase of the cytosolic concentration of dopamine, having toxic effects of its own, will also increase the availability of this catecholamine for 6-OHDA formation: establishing a potentially fatal circle. Additionally, due to the inhibition of respiration by dopamine and 6-OHDA, there may be a shortage of ATP for vesicular uptake or the neurotransmitter. Thus, it appears that the most probable target for 6-OHDA-induced damage is the pre-synaptic neuron, 6-OHDA can also be expected to occur in significant amounts in synaptic vesicles and be released alongside with other catecholamines into the synaptic cleft. This might result in its competition with dopamine for the pre- and postsynaptic receptors and the re-uptake sites on the pre-synaptic neuron. Both the receptors and the transporter may be expected to be targets for a site-specific damage by 6-OHDA and its oxidation products. The inactivation of the re-uptake transporter might bring about extracellular dopamine accumulation. Competitive binding of 6-OHDA to the metabotropic of D_1 and D_2 receptors with subsequent covalent inhibition will disturb the synaptic function by interfering with the release of the regulatory α-subunit of the G-protein. This can affect cyclic AMP formation and calcium metabolism through the activation of inositol-triphosphate formation (Kebabian, 1994). A similar effect has been demonstrated: in 6-OHDA lesioned rats, supersensitivity of the dopamine D_1 and D_1/D_2 receptors correlated with a two fold activation of CTPase (Inoue et al., 1994), but did not involve upregulation of the receptors or of G-protein mechanisms (Ueda et al., 1995). The interruption of the synaptic contact will also stop trophic support to the pre-synaptic neuron and may trigger its degeneration. Accumulation of dopamine and 6-OHDA is expected to enhance free radical formation in the extracellular space.

4. Conclusions

The finding of endogenous 6-OHDA in parkinsonian patients prompted us to highlight the potential role of this potent neurotoxin in the development of neurodegenerative disease. We pointed out a chain of toxic events starting from failure of mitochondrial respiration which eventually brings about the complete degeneration of the neuronal network and also took the morphological and biochemical findings in post-mortem brains of parkinsonian patients into account. However, as space is limited, we mainly focused on the neuronal effects of 6-OHDA. A more complete picture must also include the effects of 6-OHDA on the different types of glia.

From our data it becomes clear that free radical formation is not involved in the inhibition of mitochondrial respiration by 6-OHDA. We conclude that in neurodegenerative processes, respiratory inhibition and oxidative stress may be operative at the same time but they are not necessarily linked. 6-OHDA is easily oxidizable and can also take part in free radical reactions as well as in free radical forming reactions, like the metabolic monoamine oxidation (see Fig. 4). It is not only a respiratory toxin but also a clastogen and mutagen in salmonella and chinese hamster ovary tester strains. There is strong evidence, that these processes are free radical mediated (Gee et al., 1992). The different toxic mechanisms appear to act independantly but synergistically on the neuronal network being both responsible for the slow degeneration, which is characteristic for the human neurodegenerative diseases.

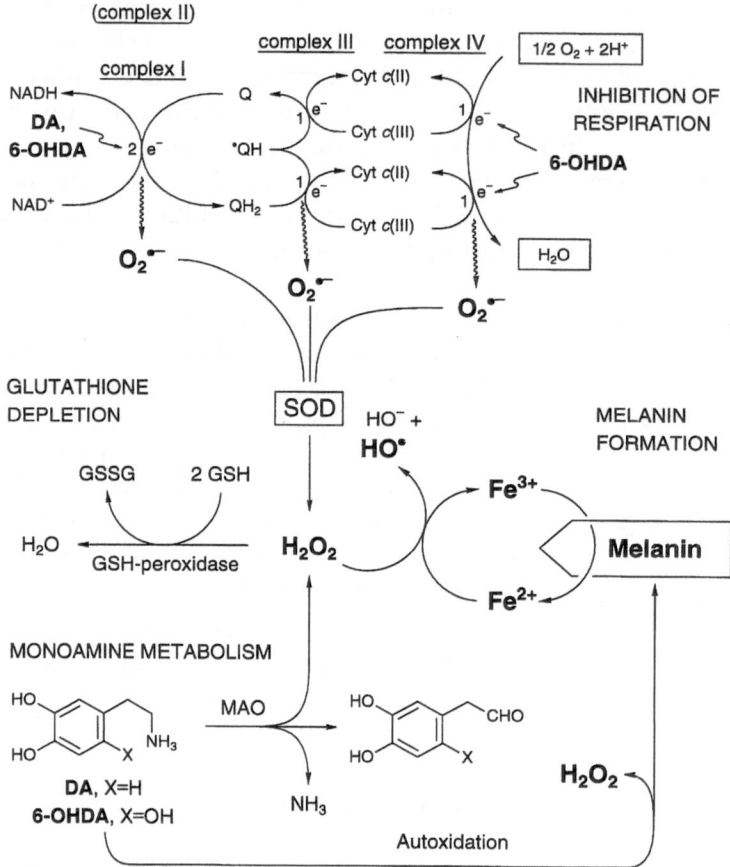

Fig. 4. The influence of 6-OHDA on cellular respiration and free radical forming processes

References

Altar CA, Boyland CB, Fritsche M, Jones BE, Jackson C, Wiegand SJ, Lindsay RM, Hyman C (1993) Efficacy of brain-derived neurotrophic factor and neurotrophin-3 on neurochemical and behavioral deficits associated with partial nigrostriatal dopamine lesions. J Neurochem 63: 1021–1032

Andrew R, Watson DG, Best SA, Midgey H, Wenlong H, Petty RKH (1993) The determination of hydroxydopamines and other trace amines in the urine of Parkinsonian patients and normal controls. Neurochem Res 18: 1175–1177

Beal MF, Hyman BT, Koroshetz W (1993) Do defects in mitochondrial energy metabolism underlie the pathology of neurodegenerative diseases? Trends Neurosci 16: 125–131

Ben-Shachar D, Youdim MBH (1990) Selectivity of melanized nigra-striatal dopamine neurons to degeneration in Parkinson's disease may depend on iron-melanin interaction. J Neural Transm [Suppl] 29: 251–258

Ben-Shachar D, Youdim MBH (1991) Intranigral iron injection induces behavioral and biochemical "parkinsonism" in rats. J Neurochem 57: 2133–2135

Ben-Shachar D, Eshel G, Finberg JPM, Youdim MBH (1991) The iron chelator desferroxamine (desferal) retards 6-hydroxydopamine-induced degeneration of nigrostriatal dopamine neurons. J Neurochem 56: 1441–1444

Ben-Shachar D, Zuk R, Glinka Y (1994) Dopamine neurotoxicity: inhibition of mitochondrial respiration. J Neurochem 64: 718–723

Cerruti C, Drian MJ, Kamenka JM, Privat A (1993) Protection by BTCP of cultured dopaminergic neurons exposed to neurotoxins. Brain Res 617: 138–142

Chadi G, Cao Y, Pettersson RF, Fuxe K (1994) Temporal and spatial increase of astroglial basic fibroblast growth factor synthesis after 6-hydroxydopamine-induced degeneration of the nigrostriatal dopamine neurons. Neuroscience 61: 891–910

Cohen G, Werner P (1994) Free radicals, oxidative stress, and neurodegeneration. In: Calne DB (ed) Neurodegenerative diseases. Saunders, Philadelphia, pp 139–162

Decker DE, Althaus JS, Buxser SE, VonVoigtlander PF, Ruppel PL (1993) Competitive irreversible inhibition of dopamine uptake by 6-hydroxydopamine. Res Commun Chem Pathol Pharmacol 79: 195–208

Dexter DT, Jenner P, Schapira AHV, Mardsen CD (1992) Alterations in level of iron, ferritin ans other trace metals in neurodegenerative disease affecting basal ganglia. Ann Neurol 32: 94–100

Erecinska M, Dagani F (1990) Relationship between the neuronal sodium/potassium pump and energy metabolism. Effects of K+, Na+, and adensine triphosphate in isolated brain synaptosomes. J Gen Physiol 95: 591–616

Filloux F, Townsend JJ (1993) Pre- and postsynaptic neurotoxic effects of dopamine demonstrated by intrastriatal injection. Exp Neurol 119: 79–88

Gee P, San RH, Davison AJ, Stich HF (1992) Clastogenic and mutagenic actions of active species generated in the 6-hydroxydopamine/oxygen reaction: effects of scavengers of active oxygen, iron, and metal chelating agents. Free Radic Res Commun 16: 1–10

Glinka Y, Youdim MBH (1995) Inhibition of mitochondrial complexes I and IV by 6-hydroxydopamine. Eur J Pharmacol Environ Toxicol Pharmacol 292: 329–332

Glinka Y, Gassen M, Youdim MBH (1996a) Iron and neurotransmitter function in brain. In: Connors JR (ed) Metals and oxidative damage in neurological disorders. Plenum Publishing Corp, New York

Glinka Y, Tipton KF, Youdim MBH (1996b) Nature of inhibition of mitochondrial respiratory complex 1 by 6-hydroxydopamine. J Neurochem (in press)

Gordon BG, DeBoer JM, Wooldridge LD, Sharp JG (1991) Effect of 6-hydroxydopamine on murine hepatopoetic stem cells: enhanced cytotoxicity on megakaryocyte colony forming units. Life Sci 49: 121–127

Inoue A, Ueda H, Nakata Y, Mizu Y (1994) Supersensitivity of quinpirole-evoked GTPase activation without changes in gene expression of D2 and Gi protein in the striatum of hemi-dopaminergic lesioned rats. Neurosci Lett 175: 107–110

Karoum F, Chrapusta SJ, Egan MF, Wyatt RJ (1993) Absence of 6-hydroxydopamine in the rat brain after treatment with stimulants and other dopaminergic agents: a mass fragmentographic study. J Neurochem 61: 1369–1375

Kebabian JW (1994) Neurotransmitter receptors in neurodegeneration. In: Calne DB (ed) Neurodegenerative diseases. Saunders, Philadelphia, pp 119–128

Kienzl E, Puchinger L, Jellinger K, Linert W, Stachelberger H, Jameson RF (1996) The role of the transition metals in the pathogenesis of Parkinson's disease. J Neurol Sci 4766

Kita T, Wagner GC, Philbert MA, King LA, Lowndes HE (1995) Effects of pargyline and pyrogallol on the methamphetamine-induced dopamine depletion. Mol Chem Neuropathol 24: 31–41

Kotlyar AB, Sled VD, Vinogradov AD (1992) Effect of Ca^{2+} ions on the slow active/inactive transition of the mitochondrial NADH-ubiquinone reductase. Biochim Biophys Acta 1098: 144–150

Mizuno Y, Ohta S, Tanaka M, Takamiya S, Suzuki K, Sato T, Oya H, Ozawa T, Kagawa Y (1989) Deficiencies in complex I subunits of the respiratory chain in Parkinson's disease. Biochem Biophys Res Commun 163: 1450–1455

Mizuno Y, Shin-Ichiro I, Hattori N, Yoshino H (1993) Mitochondria in Parkinson's disease. In: Calne DB, Horowski R, Mizuno Y, Poewe WH, Riederer P, Youdim MBH (eds) Advances in research on neurodegeneration. Birkhäuser, Boston, pp 55–66

Monterio HP, Winterbourne CC (1989) 6-Hydroxydopamine releases iron from ferritin and promotes ferritin-dependent lipid peroxidation. Biochem Pharmacol 38: 4144–4182

Morel I, Cillard J, Lescoat G, Sergent O, Pasdeloup N, Ocaktan AZ, Abdallah MA, Brissot P, Cillard P (1992) Antioxidant and free radical scavenging activities of the iron chelators pyoverdin and hydroxypyrid-4-ones in iron-loaded hepatocyte cultures: comparison of their mechanism of protection with that of desferrioxamine. Free Radic Biol Med 13: 499–508

Naveilhan P, Neveu I, Jehan F, Baudet C, Wion D, Brachet P (1994) Reactive oxygen species influence nerve growth factor synthesis in primary rat astrocytes. J Neurochem 62: 2178–2186

Offen D, Ziv I, Gorodin S, Barzilai A, Malik Z, Melamed E (1995) Dopamine-induced programmed cell death in mouse thymocytes. Biochim Biophys Acta 1268: 171–177

Orrenius S, Nicotera P (1994) The calcium ion and cell death. J Neural Transm [Suppl] 43: 1–11

Perumal AS, Gopal VB, Tordzro WK, Cooper TB, Cadet JL (1992) Vitamin E attenuates the toxic effects of 6-hydroxydopamine on free radical scavenging systems in rat brain. Brain Res Bulletin 29: 699–701

Sampath D, Jackson GR, Werrbach-Perez K, Perez-Polo JR (1994) Effects of nerve growth factor on glutathione peroxidase and catalase in PC12 cells. J Neurochem 62: 2476–2479

Schwartz JP, Nishiyama N (1994) Neurotrophic factor gene expression in astrocytes during development and following injury. Brain Res Bull 35: 403–407

Spina MB, Squinto SP, Miller J, Lindsay RM, Hyman C (1992) Brain-derived neurotrophic factor protects dopamine neurons against 6-hydroxydopamine and N-methyl-4-phenylpyridinium ion toxicity: involvement of glutathione system. J Neurochem 59: 99–106

Stadtman ER (1993) Oxidation of free amino acids and amino acid residues in proteins by radiolysis and by metal-catalyzed reactions. Ann Rev Biochem 62: 797–821

Ueda H, Sato K, Okumura F, Inoue A, Nakata Y, Yue JL, Mizu Y (1995) Supersensitization of neurochemical responses by L-DOPA and dopamine receptor agonists in the striatum of experimental Parkinson's disease model rats. Biomed Pharmacother 49: 169–177

Vroegop SM, Decker DE, Buxser SE (1995) Localization of damage induced by reactive oxygen species in cultured cells. Free Radic Biol Med 18: 141–151

Yokoyama C, Okamura H, Ibata Y (1993) Resistance of hypothalamic dopaminergic neurons to neonatal 6-hydroxydopamine toxicity. Brain Res Bulletin 30: 551–559

Youdim MBH, Riederer P (1993) The role of iron in senescence of dopaminergic neurons in Parkinson's disease. J Neural Transm [Suppl] 40: 57–67

Authors' address: Prof. Dr. M.B.H. Youdim, Department of Pharmacology, Bruce Rappaport Faculty of Medicine, Technion, P.O.B. 9649, Haifa 31096, Israel.

Induction of mitosis-related genes during dopamine-triggered apoptosis in sympathetic neurons

A. Shirvan[1], **I. Ziv**[1], **A. Barzilai**[2], **R. Djaldeti**[1], **R. Zilkh-Falb**[2], **T. Michlin**[1], and **E. Melamed**[1]

[1]Felsenstein Medical Research Center and Neurology Department, Rabin Medical Center (Beilinson Campus), Petah Tikva, [2]Department of Nerobiochemistry, Tel Aviv University, Tel Aviv, Israel

Summary. It was suggested that neuronal degeneration in Parkinson's Disease (PD) is linked to dopamine (DA) toxicity. Dopamine has been shown to induce programmed cell death in both neuronal and non-neuronal cell types. We examined the molecular changes associated with dopamine-triggered apoptosis in sympathetic neurons using the differential display approach, and isolated 14 different DA responsive genes whose expression is altered during the early stages of the apoptotic process. Nine of these genes are upregulated and five are downregulated in response to DA exposure. Two of the upregulated genes were identified as cyclin B2 and a chicken homologue of chaperonin, a member of the heat shock protein family. Total increase in mRNA expression of both genes after 12 hours of exposure to DA was 40%. These two genes participate in cell cycle control and are specifically involved in determining entry of dividing cells into mitosis. Upregulation of mitosis- related genes in postmitotic sympathetic neurons undergoing apoptosis, may be indicative of an abortive attempt of these neurons to re-enter the cell cycle prior to their death. Possible implications to neuronal degeneration in PD are discussed.

Introduction

PD is a progressive neurodegenerative disease, in which the dopaminergic neurons in the substantia nigra degenerate slowly. Although it has been the subject of intensive research, the etiology of this nigral neuronal loss is still enigmatic. A current hypothesis suggests that this process is associated with excessive oxidant stress, possibly induced by DA (Olanow, 1993). Until recently, nigral neuronal degeneration in PD was considered to be of the necrotic type, caused by lipid peroxidation and rupture of cellular membranes (Olanow, 1993). It was assumed that the driving force of such a process was the generation of toxic free radicals during the metabolism of DA in the dopaminergic neurons. However, such mechanism would be expected to cause rapid nigral destruction by cell lysis. In contrast, histopathology of nigral degeneration in PD is characterized by a protracted death process and lack of obvious inflammatory reaction. Also, some of the remaining neurons show condensed heterochromatin and appear shrunken. These observations maybe more consistent with an apoptotic mode of cell death. Apoptosis is a specific program of genetically encoded events that depends on active metabolism and leads to the cell's own destruction by a typical chain of events. It has a major physiological role in the differentiation and organization of the nervous system. During the development of the neuron, such death program can be easily activated, enabling proper organization and maturation (Baringa, 1993). However, it maybe assumed that in the fully differentiated neurons, this program is inactivated and kept in a restrained form, in order to allow for longevity of the neuron. However, it is likely that inappropriate activation of programmed cell death is involved in pathogenesis of degenerative neurological disorders (Thompson, 1995; Steller, 1995).

Recently, we reported that DA, the endogenous neurotransmitter of nigrostriatal neurons, is capable of initiating apoptosis in neuronal and non-neuronal cell types from either primary cultures or cell lines (Ziv et al., 1994; Offen et al., 1995; Offen et al., 1996), and that post mitotic chick sympathetic neurons can be used as a model system for studying the toxic effects of DA (Ziv et al., 1994; Zilkha-Falb et al., 1996).

Exposure of these cells to DA initiated the characteristic apoptotic cascade, as evaluated by a wide array of methods (Ziv et al., 1994; Zilkha-Falb et al., 1996).

We have also shown that DA toxicity could be inhibited by overexpression of Bcl-2 in PC-12 cells (Ziv et al., 1996; Offen et al., 1996), suggesting a protective role for Bcl-2 in preventing cells from oxidative stress-induced apoptosis.

To better understand the mechanisms that control neuronal cell apoptosis induced by DA, we initiated comparative studies aimed towards identifying possible genes involved in this process. For that purpose we employed the recently-described differential display technique (Liang and Pardee, 1992; Liang et al., 1993) to sympathetic neurons, and compared gene expression in DA-treated and untreated cells. The major advantage of that approach is the ability to isolate, in an unbiased way, a whole set of genes that are associated with the death process. Using this technique we have isolated several genes that are transcriptionally regulated during DA-induced apoptosis in neuronal cells. In particular, we report the induction of mitosis-related genes, such as cyclin B2 and chaperonin. Our results provide evidence that the induction of both cyclin B2 and chaperonin coincides with the active death process. The induced expression of cell cycle-related genes in post-mitotic neurons may therefore reflect an abortive attempt of neurons to re-enter the cell cycle, resulting in neuronal apoptosis.

Materials and methods

Cells growth and treatments

Primary cultures of sympathetic neurons were prepared essentially as described previously (Ziv et al., 1994), except that antimitotic drugs (used to inhibit non neuronal dividing cells) such as fluoredeoxyuridine and uridine were used in a concentrations of 20 μM each.

Cells were plated at a density of 10^6 cells/35 mm dish, and grown in culture for 4 days. At that time dopamine was administered at a concentration of 300 µM. Cell viability was assessed by the trypan blue assay, as described previously (Ziv et al., 1994).

RNA preparation

Total RNA was extracted from sympathetic neurons using Triazol reagent from BRL, and treated by DNase I (BRL), according to manufacturer's instructions. RNA was dissolved in Diethyl Pyrocarbonate (DEPC) treated water at a concentration of 100 ng/ml and analyzed in agarose gel electrophoresis to assess integrity prior to use in the differential display reactions.

Differential display

Differential display protocol was essentially as previously described (Liang et al., 1993). For each reaction 50–100 ng of total RNA was used. The primers utilized for the reactions were as follows: 3 antisense primers: $dT_{12}CA$; $dT_{12}AG$ and $dT_{12}GC$. 4 sense primers were P1: 5'-TGTCAGTCAG; P2: 5'GGTACTAAGC; P3: 5'TCGATACAGG and P4: 5'-ACAGTTGACA. Reaction conditions were as follows: for reverse transcription, incubation was at 42°C for 90 minutes and 95°C for 5 minutes. For PCR amplification: 40 cycles of 30 seconds at 94°C, 1 minute at 42°C and 30 seconds at 72°C. Reactions were carried out in triplicates and run on 6% acrylamide sequencing gels.

Cloning and sequencing of bands of interest

cDNA bands of interest were cut out from the dried gel, and eluted with water for 60 minutes, at room temperature. DNA was put directly into a second PCR reaction, with the same arbitrary primers used in the Differential Display reaction. Reaction conditions were the same except that dNTP concentration was 20 µM. Reamplified bands were analyzed on 4% agarose gels and extracted using the GETsorb kit from Genomed Inc. N.C. DNA was cloned into pGEM-T vector (Promega), and sequencing was performed using Sequnase II from Amersham. DNA sequences were analysed using the Blast software (Benson et al., 1993), or the FASTA program (GCC software, Madison W1).

Northern blot analysis

5 µg of total RNA were run on each lane (corresponding to 10^6 neurons) either treated or untreated by DA. Northern blot was carried out as described (Sambrook et al., 1989) and Qiabrane nylon membranes were used (Qiagen). RNA was crosslinked by UV.

Probe preparation

cDNA of interest were labeled with 32p-dCTP by random priming, using Rediprime kit from Amersham, and 5×10^7 cpm were used for hybridization. Blots were washed at room temperature with $0.1 \times$ SSC; 0.1% SDS. Bands intensity was quantitated by soft laser scanning densitometer (Biomed Instruments) and by B.I.S. 202 D gel documentation system (Dinco and Renium). As a control for loading and transfer, membranes were re-hybridized to an oligonucleotide of 18S ribosomal RNA which was kinased according to

(Sambrook et al., 1989). RNA quantities in each lane were normalized against ribosomal 18S rRNA level. As an internal control, expression of GAPDH (Glyceraldehyde 3-phosphate dehydrogenase) was monitored by rehybridizing the same blots with a probe of Rat GAPDH (obtained from Dr. D. Lazar).

Results

Determination of early stages of the apoptotic process

To identify possible genes involved in apoptotic neuronal degeneration, we used the model of DA-induced neuronal death, and followed this process in primary cultures of chick postmitotic sympathetic neurons. Upon applying DA at a concentration of 300 μM to the cultures, a rapid death process was observed (Fig. 1). Exposure of the culture to DA for 13–14 hours resulted in a decrease of 50% in cell viability, as judged by trypan blue assay. Ten hours later, at the 24 hours time point, all cells were dead.

We were interested in the early stages of the apoptotic process before the common pathway to programmed cell death is initiated and therefore concentrated on the first 12 hours of DA exposure, in which 80% of the cells are still viable. At this phase, the cellular processes that precede neuronal degeneration has probably started and could be studied. Removal of the apoptotic insult 6 hours after its administration by replacing the DA containing medium with a fresh medium without DA, could rescue the cells from death, and they continued to live in culture for at least 24 hours. However, removal of DA from the cultures 12 hours following its application, resulted in 21% survival (Fig. 2). Therefore, the time points between 6–12 hours are the phases in which cells are committed to die. In order to follow the molecular events

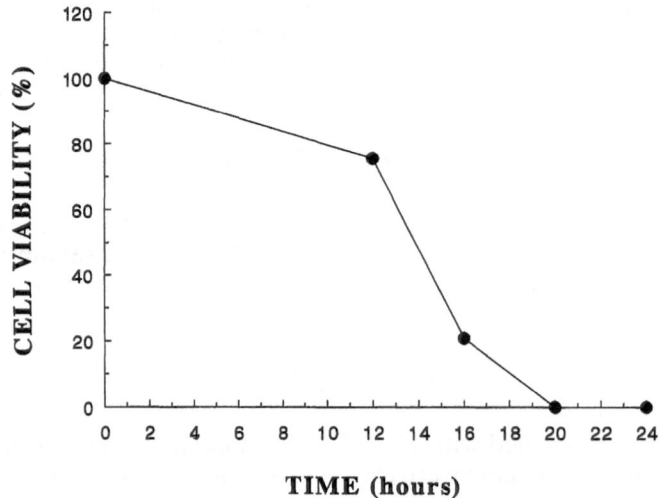

Fig. 1. Time course of DA induced apoptosis in sympathetic neurons. Neurons were treated by 300 μM of DA for various time periods and their survival in culture was monitored using trypan blue assay. 50% death is observed after 14 hours of DA exposure

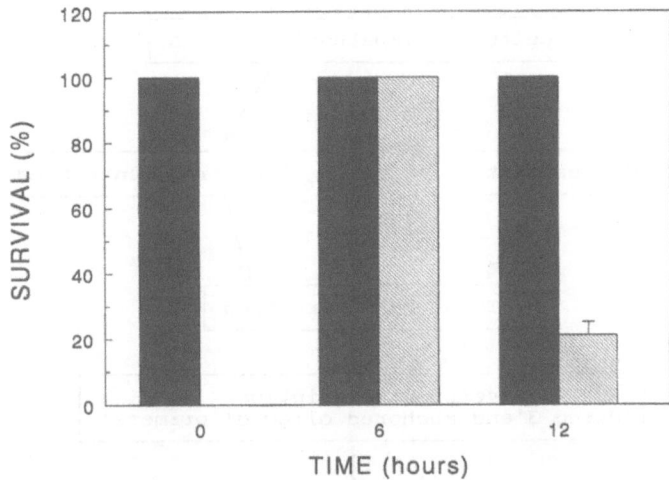

Fig. 2. Rescue of cells from dopamine induced death by DA removal. Neurons were treated with DA (300 μM) (black bars). DA was washed out from cultures and replaced with fresh medium, at 6 and 12 hours, respectively (grey bars). Cells were allowed to grow in culture and their survival was monitored 24 hours after the onset of the DA induced death process. 6 hours of DA exposure do not affect cell survival and 12 hours of exposure reduced survival rate by 80%

associated with the active death process, we chose to compare cells that were treated by DA with untreated cells, 12 hours following exposure.

Selective expression of apoptosis-related genes

Potential mediators of the DA-induced apoptotic process were isolated using the Differential Display method, capable of detecting differences in the mRNA repertoire between various cell populations. Schematic representation of the experimental approach is presented in Fig. 3. DA-treated and control untreated cells were used for total RNA preparation. RNA primed by poly dT primers anchored to two additional bases at their 3' ends, was used for reverse transcription. PCR amplification was followed using arbitrary primers. We have used a combination of 3 antisense and 4 sense primers, and carried out the Differential Display reactions. An example of 3 different reactions is presented in Fig. 4. Among the approximately 100 bands displayed in every reaction, each representing cDNA molecules, multiple bands that showed differential expression can be detected and several of them are indicated by arrows. Some of the cDNAs were upregulated in the presence of DA and some were downregulated. The reproducibility of the altered expression was validated by carrying out the same reaction from three different independent RNA preparations. We chose 14 different cDNA bands, ranging in size from 200–400 base pairs for further characterization. Each of these bands was rescued from the dried gel, eluted, and used for a second amplification reaction. The reaction product was run on agarose gels, and bands were isolated as described in Materials and Methods.

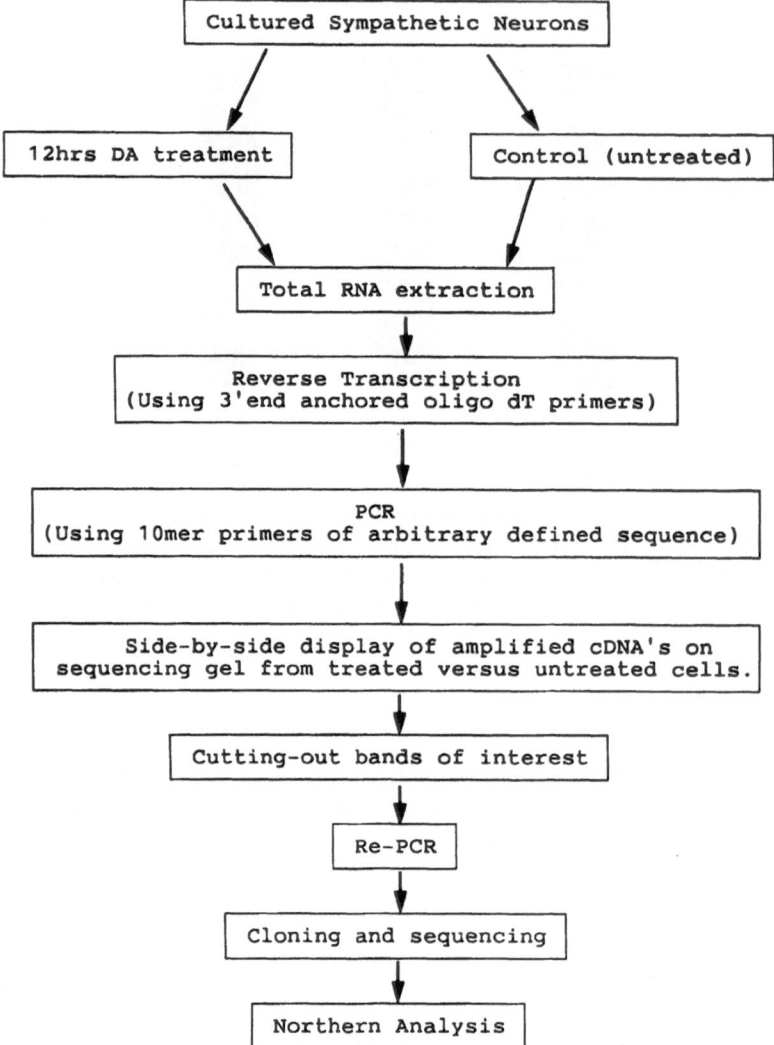

Fig. 3. Schematic representation of the experimental approach. Each step is described in detail in the text

Identification of DA responsive genes

All the 14 cDNA bands isolated from the differential display reactions were subcloned into a pGEM-T vector. Existence of the appropriate size band was verified by restriction enzyme analysis. As shown in Fig. 5, each one of the clones was cut by the enzymes SacII and SpeI, that flank the ligation site into the vector, thus releasing the insert which is then separated in a 4% agarose gels. 12 different clones can be seen in Fig. 5, ranging in size from 200–400 base pairs. Each of the clones was subject to sequence analysis, and compared with

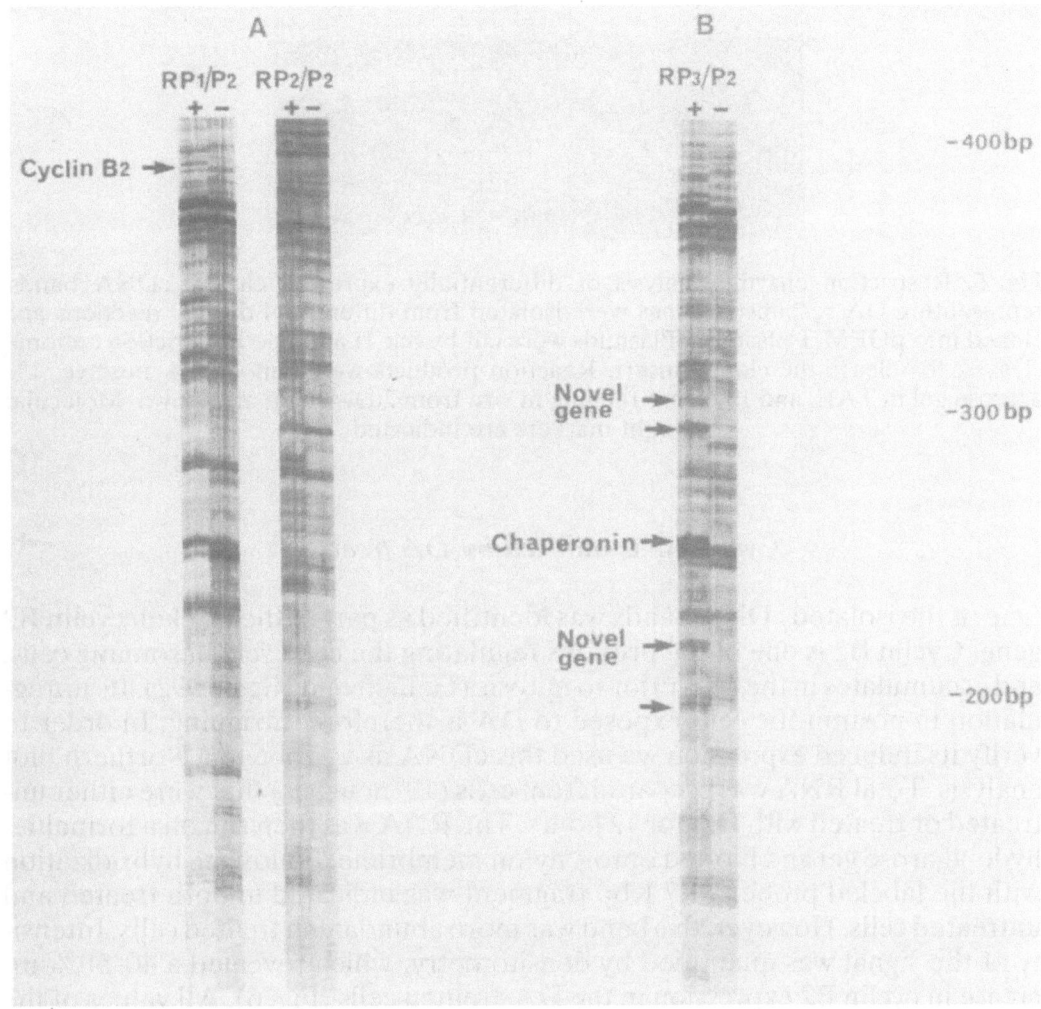

Fig. 4. Differential display reactions. mRNA from chick sympathetic neurons, differentially expressed as a result of DA treatment is presented. Three different reactions are shown using 3 different combinations of random primers (each reaction was repeated 3 times with various RNA preparations and expression pattern of mRNA was confirmed). Samples were separated on polyacrylamide sequencing gel. Arrows indicate changes in gene expression following DA exposure. Specific bands identified as encoding for cyclin B2 and chaperonin are marked. DNA size markers are shown

updated DNA and protein data bases. It was found that the different clones represent 14 distinct genes, 11 of which do not show any significant homology to previously characterized genes. We believe that they represent novel genes that are involved in DA induced apoptotic death. Nine of the 14 clones were induced by DA and 5 were reduced. Two of the clones that showed induced expression, following 12 hours of DA treatment, were found to be homologues to chicken cyclin B2 (99% homology on the DNA and protein level) and to a mouse T-complex protein (75% homology on the DNA and 91% homology on the protein level).

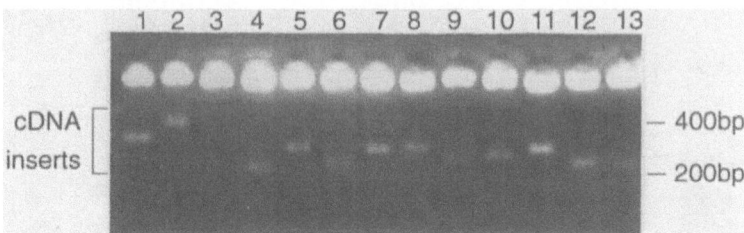

Fig. 5. Restriction enzyme analysis of differentially expressed clones. cDNA bands, representing DA responsive genes were isolated from differential display reactions and cloned into pGEM-T plasmids. Plasmids were cut by Sac II and Spe I restriction endonucleases, to release the cloned insert. Reaction products were run on 3% nusieve, 1% agarose gel in TAE, and 13 clones ranging in size from 200–400 bp are shown. Molecular weight markers are indicated

Cyclin B2 is induced by DA treatment

One of the isolated cDNA bands was identified as part of the chicken cyclin B2 gene. Cyclin B2 is one of the proteins regulating the cell cycle in somatic cells, and accumulates in the cells prior to mitosis (Gallant and Nigg, 1992). Its upregulation in postmitotic cells exposed to DA is therefore intriguing. In order to verify its induced expression we used this cDNA as a probe in a Northern blot analysis. Total RNA was prepared from cells (10^6 neurons) that were either untreated or treated with DA for 12 hours. The RNA was then run on a formaldehyde-agarose gel and blotted onto a nylon membrane. Following hybridization with the labeled probe, a 1.7 Kbp fragment was indicated in both treated and untreated cells. However, this band was more abundant in treated cells. Intensity of the signal was quantified by densitometry, which revealed a 40–50% increase in cyclin B2 expression in the DA-treated cells (Fig. 6). All values of the RNA levels were normalized according to 18S rRNA levels, which were monitored by hybridizing the same membrane with the appropriate 18S rRNA probe. As a control, expression of GAPDH was monitored by re-hybridizing the same membrane with a probe of GAPDH. We observed no difference in GAPDH expression in both treated and untreated cells. Upregulation of cyclin B2 correlated with the death process, since it was observed at 12 hours following DA treatment, a time point in which the neurons are already committed to die. This process however, precedes any morphological changes that are typical for apoptotsis, suggesting that cell cycle related events are involved in the very early stages of neuronal apoptosis.

Chaperonin mRNA induction in DA-treated cells

Another gene that showed altered levels in the differential display experiments, was the chicken homologue of the mouse T-complex protein. This protein is a member of the chaperonin family of heat shock proteins (Willison et al., 1986), also known to be induced by oxidative stress. The cDNA

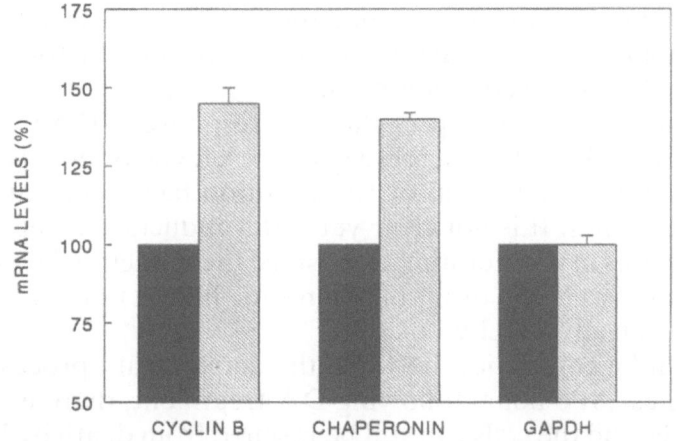

Fig. 6. DA induced expression of cyclin B2 and chaperonin. Sympathetic neurons were exposed to DA for 12 hours and total RNA was prepared and analyzed by Northern blot (grey bars), in comparison to RNA from untreated cells (black bars). Cyclin B2 and chaperonin cDNA were used as probes, as well as a probe of GAPDH. Diagram was calculated by computerized densitometry of the exposed x-ray film. The quantitative results are presented as the ratio of cyclin B2, chaperonin or GAPDH signal, to the 18S rRNA signal. 40% increase in cyclin B and chaperonin expression was observed. Each experiment was repeated 3 times with different RNA preparations

encoding the gene is indicated by an arrow in Fig. 4. Northern blot analysis revealed an elevated expression as a function of DA treatment (Fig. 6). A total increase of 40% above the control level of untreated cells, was observed 12 hours after exposure to DA. The level of increase in mRNA expression of chaperonin is of the same order of magnitude as that of cycline B2, suggesting that both might be regulated by similar mechanisms.

Discussion

DA induces programmed cell death in chick sympathetic neurons, by a process that is characterized by a wide array of ultrastructural changes, typical of cells undergoing apoptosis (Ziv et al., 1994). The purpose of this work was to investigate molecular events associated with DA toxicity, using the Differential Display approach (Fig. 3). We identified a large number of cDNA molecules, whose expression was consistently altered as a function of DA treatment (DA-responsive genes) (Fig. 4). Upon cloning and sequencing 14 of these cDNA molecules (Fig. 5), 11 were found to encode novel genes. Since expression of these genes is transcriptionally regulated at the early stages of the apoptotic process, before any morphological sign of PCD are observed, we postulate that some of them are involved in dysregulation of the control mechanism of apoptosis (i.e. they might be part of the cellular events that determine the fate of the neurons towards self destruction versus survival). These early events that precede neuronal degeneration in cultured cells might also take place in neurons that are sick and destined to die. The large number of DA responsive genes, which exceeded the 14 clones we isolated, is consistent with a possibility of transactivation of a transcription factor, whose action is modulated by DA-induced oxidative stress.

Two of the genes that were upregulated in response to DA treatment were cyclin B2 and chaperonin. The increase in mRNA level for both was 40%. This could imply a similar mechanism or transcription factor regulating the transcription of both genes. It is not clear yet if this induction is regulated by the same mechanisms as in proliferating cells, since the induction level observed in preparation to mitosis is one order of magnitude higher than the induction we observed in neurons treated by DA.

We also found a correlation between the active death process and induction of both genes. At 6 hours following DA treatment, there is no change in expression levels, and the cells can still be rescued from death by DA removal. However, at 12 hours when the cells are already committed to die, they overexpress both genes (Fig. 2 and Fig. 6).

Cyclin B2 is a key protein, regulating entry into the mitosis stage in proliferating cells. Both its mRNA and protein exhibit cycles of synthesis and destruction in correlation to the onset of mitosis (Pines and Hunter, 1989, 1990; King, 1994).

Cyclin B2 levels increase dramatically in preparation for mitosis, forming a complex with its catalytic subunit cdc^{p34}. At the anaphase stage of mitosis, cyclin B2 protein is rapidly degraded (Murray, 1995). This process has been well established for dividing cells. By contrast, sympathetic neurons are terminally differentiated and postmitotic, and have exited from the cell cycle irreversibly. Therefore their induction during the death process may hold some hints for the possible responsible mechanism.

Our observation that a member of the heat shock family of proteins is accumulated during the apoptotic process is intriguing, since they were recently implicated in participating in cell cycle control. In yeasts, HSP90 was found to be a positive regulator of Wee1 kinase (Aligue et al., 1994) which phosphorylates the complex of cyclin B2 and $cdc2^{p34}$. It is this phosphorylation that inactivates the complex and prevents the actual start of mitosis (Heald et al., 1993). Entry into mitosis requires dephosphorylation of this complex, and therefore HSP90 protein levels can dynamically regulate the initiation of mitosis. It is expected that during mitosis, HSP levels will be lower and accumulation of HSP within the cells, will raise the amount of phosphorylated complexes, thus preventing entry into mitosis. In the postmitotic neurons exposed to DA, two contradictory signals occur: an accumulation of cyclin B2, signaling the cells that mitosis is about to commence, and accumulation of the heat shock protein that may prevent activation of cyclin B-cdc2 complex, thereby halting the mitotic process. We propose that such deregulated expression of mitosis-associated proteins, happening in postmitotic neurons, leads them towards their self destruction. However, further experiments are needed to unravel the detailed nature of this contradictory signals and the interaction between the chick chaperonin and phosphorylation of cyclin B-cdc2 complex.

An aberrant process of mitosis has been shown to occur when cell cycle components are overexpressed or sustain mutations (Russel and Nurse, 1987;

Heald et al., 1993). In such cases, normal mitotic checkpoints are bypassed, releasing tight control on the order of cell cycle events (Russel and Nurse, 1987). Some of the morphological changes taking place in normal mitosis are also displayed in the abortive mitosis, including spindle formation (Ucker, 1991). However multiple spindles are often observed, and chromosomes fail to orient properly (Heald et al., 1993). It is of interest that ectopic expression of either one of the proteins, cyclin B2 and TCP-1 in somatic cells, leads to multiple spindle formation and unsuccessful mitosis (Gallant and Nigg, 1992; Ursic and Culberston, 1991).

The body of evidence that links apoptosis to cell cycle in neuronal and non neuronal cells is growing rapidly (Reviewed in: Rubin et al., 1994; King and Cidlowski, 1995; Bresden, 1995). The morphology of cells undergoing apoptosis is similar in general to that of cells undergoing both normal mitosis and an aberrant form of mitosis called mitosis catastrophe. During each of these processes, cells release substrate attachment, loose volume, condense their chromatin and disassemble nuclear lamina. Cell cycle deregulation has been shown to induce mitosis catastrophe and may be involved in triggering apoptosis (Rubin et al., 1994). Evidence exists for induction of other cell cycle related genes during apoptosis (Bredsen, 1995). Of particular interest is cyclin D1 which is upregulated in rat sympathetic neurons following NGF deprivation (Freemen et al., 1994). By contrast, we were unable to detect induction of cyclin D1 in our model (not shown). We suggest that although NGF deprivation and oxidative stress are capable of inducing apoptosis in sympathetic neurons, the routes leading to this process can be variable.

The exact relationship between mitosis and apoptosis are still poorly understood, but in view of the presented findings, we speculate that the processes taking place in DA-treated postmitotic neurons may also occur during the course of nigral degeneration in PD. Dysregulated induction of cell cycle related genes can lead the neuron into a mitotic catastrophe, which may explain the histopathology of sick neurons in PD.

Acknowledgement

The research was supported in part by Teva Pharmaceutical Industries Ltd., and the National Parkinson Foundation, Miami, Florida, USA.

References

Aligue R, Akhavan-Niak H, Russell P (1994) A role for Hsp90 in cell cycle control: Wee1 tyrosine kinase activity requires interaction with Hsp90. EMBO Journal 13 (24): 6099–6106

Barinaga M (1993) Death gives birth to the nervous system, but how? Science 259 (5096): 762–763

Benson D, Lipman DJ, Ostell J (1993) Gene Bank Nucleic Acid Res 21: 2963–2965

Boobis AR, Fawthrop DJ, Davies DS (1989) Mechanisms of cell death. Trends Pharmacol Sci 10 (7): 275–280

Bredesen DE (1995) Neural apoptosis. Ann Neurol 38: 839–851

Freeman RS, Estus S, Johnson EM (1994) Analysis of cell cycle related gene expression in postmitotic neurons: selective induction of cyclin D1 during programmed cell death. Neuron 12: 343–355

Gallant P Niqq A (1992) Cyclin B2 undergoes cell cycle dependent nuclear translocation and when expressed as non-destructible mutant causes mitotic arrest in HeLa cells. J Cell Biol 117 (1): 213–224

Heald R, McLoughlin M, McKeon F (1993) Human Wee1 maintains mitotic timing by protecting the nucleus from cytoplasmically activated Cdc2 kinase. Cell 74: 463–474

King KL, Cidlwoski JA (1995) Cell cycle and apoptosis: Common pathways for life and death. J Cellular Biochemistry 58: 175–180

King RW, Jackson P, Kirschner MW (1994) Mitosis in transition. Cell 79: 563–571

Liang P, Averboukh L, Pardee AB (1993) Distribution and cloning of eukaryotic mRNA by means of differential display: refinements and optimization. Nucleic Acids Res 21(14): 3269–3275

Liang P, Pardee AB (1992) Differential display of eukaryotic messenger RNA by means of the polymerase chain reaction. Science 257 (5072): 967–971

Murray A (1995) Cyclin ubiquitination: The destructive end of mitosis. Cell 81: 149–152

Offen D, Ziv I, Gorodin S, Barzilai A, Malik Z, Melamed E (1995) Dopamine-induced programmed cell death in mouse thymocytes. Biochemica Biophysica Acta 1268: 171–177

Offen D, Ziv I, Panet H, Wasserman L, Stein R, Melamed E, Barzilai A (1996) Dopamine-induced apoptosis is inhibited in PC12 cells expressing Bcl-2: evidence that the proto-oncogene may act as an antioxidant. Mol Cell Neurobiol (in press)

Olanow CW (1993) A radical hypothesis for neurodegeneration. Trends Neuroscience 16: 439–444

Pines J, Hunter T (1989) Isolation of human cyclin cDNA: Evidence for cyclin mRNA and protein regulation in the cell cycle and for interaction with p34^{cdc2}. Cell 58: 833–846

Pines J, Hunter T (1990) Human cyclin A is adenovirus E1A associated protein p60 and behaves differently from cyclin B. Nature 346: 760–763

Rubin LL, Gatchalian CL, Rimon G, Brooks SF (1994) The molecular mechanisms of neuronal apoptosis. Current Opinion in Neurobiology 4: 696–702

Russel P, Nurse P (1987) Negative regulation of mitosis by wee1, a gene encoding a protein kinase homolog. Cell 49: 559–567

Sambook J, Fritch EF, Maniatis R (1989) Molecular Cloning: A Laboratory Manual. 2nd edition. Cold Spring Harbour Laboratory Press

Steller H (1995) Mechanisms and genes of cellular suicide. Science 267: 1445–1449

Thompson CB (1995) Apoptosis in the pathogenesis and treatment of disease. Science 267 (5203): 1456–1462

Ucker CS (1991) Death by suicide: one way to go in mammalian cellular development. New Biol 3: 103–109

Ursic D, Culberston MR (1991) The yeast homolog to mouse Tcp-1 affects microtubule-associated process. Mol Cell Biol 11 (5): 2629–2640

Willison KR, Dudley K, Potter J (1986) Molecular cloning and sequence analysis of a haploid expressed gene encoding t-complex polypeptide 1. Cell 44: 727–738

Zilkha-Falb R, Ziv I, Nardi N, Offen D, Melamed E, Barzilai A (1996) Monoamines induced apoptotic neuronal cells death. Cell Mol Neurobiol (in press)

Ziv I, Melamed E, Nardi N, Luria D, Achiron A, Offen D, Barzilai A (1994) Dopamine induces apoptosis-like cell death in cultured symptathtic neurons-A possible novel pathogenetic mechanism in Parkinson's disease. Neurosci Lett 170: 136–140

Ziv I, Offen D, Haviv R, Stein R, Achiron A, Panet H, Barzilai A, Melamed E (1996) The proto-oncogene Bcl-2 inhibits cellular toxicity of dopamine: Possible implications for Parkinson's disease (submitted for publication)

Authors' address: Dr. A. Shirvan, Felsenstein Medical Research Center and Neurology Department, Rabin Medical Center (Beilinson Campus), Petah Tikva, 49100, Israel.

Neuronal vulnerability in Parkinson's disease

Etienne C. Hirsch, B. Faucheux, P. Damier, A. Mouatt-Prigent,
and **Y. Agid**

INSERM U289, Physiopathologie et Pathogenèse des Maladies Neurodégénératives,
Hôpital de la Salpêtrière, Paris, France

Summary. Although Parkinson's disease is characterized by a loss of dopaminergic neurons in the substantia nigra not all dopaminergic neurons degenerate in this disease. This suggests that some specific factors make subpopulations of dopaminergic neurons more susceptible to the disease. Here, we show that the most vulnerable neurons are particularly sensitive to oxidative stress and rise in intracellular calcium concentrations. Because both events seem to occur in Parkinson's disease this may explain why some dopaminergic neurons degenerate and other do not.

Which dopaminergic neurons degenerate in Parkinson's disease?

The main anatomobiochemical characteristic of Parkinson's disease is the loss of dopaminergic neurons in the mesencephalon. Degeneration of dopaminergic neurons is, however, heterogeneous in this structure, the degree of neuronal loss varying considerably from one catecholaminergic cell group to another. It is severe in the substantia nigra pars compacta (76% loss), intermediate in the substantia nigra pars lateralis (34% loss), ventral tegmental area (55% loss) and catecholaminergic cell group A8 (31% loss) but almost nil in the central gray substance (7% loss) (Hirsch et al., 1988).

Besides these inter-regional differences in the degree of neuronal loss in the mesencephalon, the loss of dopaminergic neurons has also been found to be variable within subgroups of neurons in the substantia nigra pars compacta. Indeed, in 1937, Hassler published a very precise description of the human substantia nigra, in which he subdivided it into more than thirty subgroups of neurons according to cytoarchitectonic and cell density criteria (Hassler, 1937). In Parkinson's disease, he found that the neuronal loss was the most prominent in the caudal and ventro-lateral part of the substantia nigra pars compacta (Hassler, 1938). More recently, Fearnley and Lees (1991) subdivided the structure into six main clusters of melanized dopaminergic neurons at a level of the substantia nigra pars compacta including the emerging fibers of the third cranial nerve. They found that the neuronal loss was massive in the ventro-lateral part of the structure and less pronounced in its dorsal part. Nevertheless, the precise identification of the dopaminergic neurons that degenerate in Parkinson's disease requires further study for at least two reasons.

1. In most of the previous studies only melanized neurons were analyzed. Although the melanized neurons are dopaminergic in the substantia nigra (Bogert et al., 1983), they represent only a subpopulation of dopaminergic neurons, since 16% of all nigral dopaminergic neurons are devoid of neuromelanin (Hirsch et al., 1988).

2. The subdivision of nigral cell groups in the previous studies was essentially based on the dopaminergic neurons which degenerate during the pathological process. However, the use of limits which are dependent on the pathological process to identify the regions which degenerate in the diseased state is open to criticism. Indeed, the absence of neurons in the ventral portion of the substantia nigra may be due to a global atrophy of the substantia nigra or to a real and selective loss of the dopaminergic neurons in the ventral part of the structure. In order to circumvent these methodological drawbacks, we recently performed a precise analysis of the distribution of neuronal loss in the parkinsonian substantia nigra using calbindin immunoreactivity as a marker which is independent of the pathological process (Damier et al., 1996).

Although, calbindin immunoreactivity has been observed in a few nigral dopaminergic neurons mainly located in the dorsal part of the structure (Hirsch et al., 1992), most of the staining consisted of a dense fiber network covering the whole substantia nigra. This intensely stained neuropil was heterogeneous, showing pockets of lower staining intensity (Fig. 1). Interestingly, a great proportion of the dopaminergic neurons, identified by their neuromelanin content, were clustered together within the zone of calbindin-poor immunostaining. Five main calbindin-poor zones were identified in the human mesencephalon, which displayed a reproducible pattern of distribution across the different subjects analyzed. This heterogeneous pattern of calbindin neuropil was preserved in Parkinson's disease in agreement with the data of Ito et al. (1992). Using this definition of the dopaminergic neurons in the substantia nigra, which is thus independent of the pathological process, the neuronal loss was found to be more pronounced in the ventral substantia nigra (97% loss) than in its dorsal part (57% loss) (Damier et al., 1996). Moreover, the neuronal loss was also more severe in the calbindin-poor zones (95% loss) than in the surrounding calbindin-rich neuropil (80% loss). Furthermore, the neuronal loss was the most intense in the largest calbindin-poor zone (98% loss). This calbindin-poor zone was located in the ventral tier of the substantia nigra rostrally and was more lateral and dorsal, caudally.

Taken as a whole these data show that Parkinson's disease is not a generalized disease of dopaminergic neurons and that some dopaminergic neurons are more susceptible to degeneration than others. Identification of the factors that render some dopaminergic neurons more vulnerable may thus provide new clues to the pathogenesis of the disease. Some of the phenotyp-

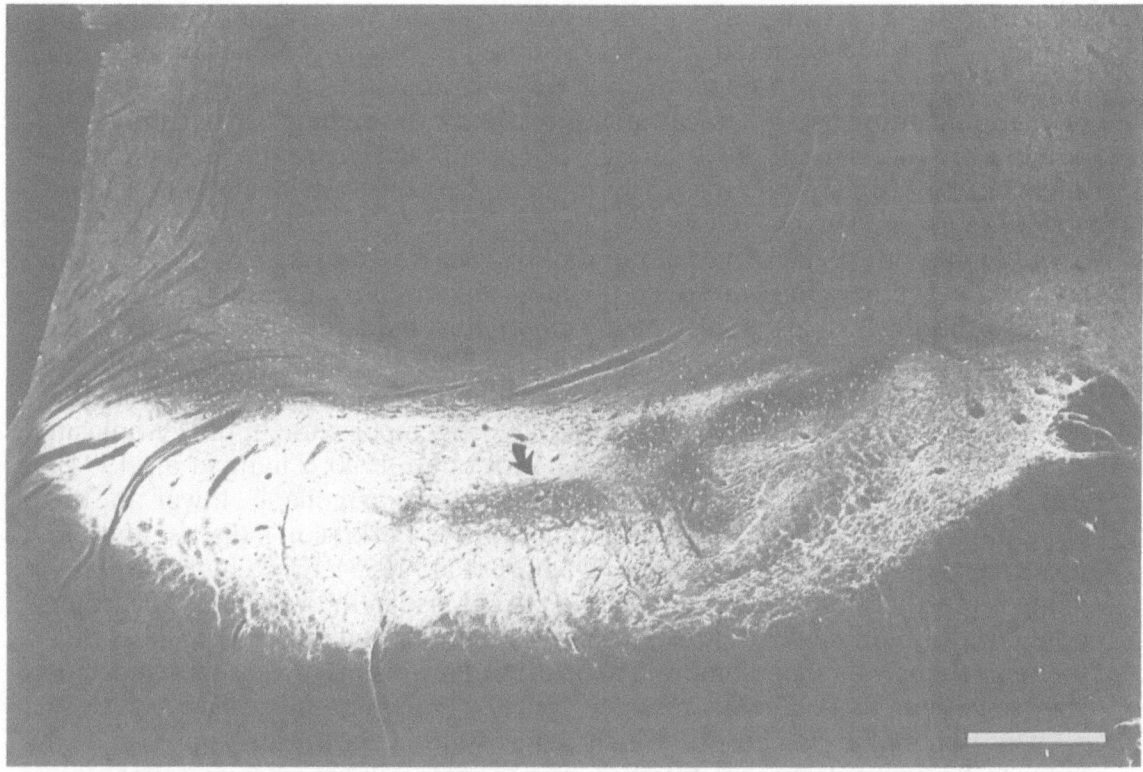

Fig. 1. Reverse contrast photograph of calbindin immunoreactivity in the mesencephalon of a control subject. Asterisk indicates a pocket of low calbindin immunoreactivity containing numerous melanized dopaminergic neurons (small white dots). Bar indicates 2 mm

ical differences between the dopaminergic neurons that are vulnerable in Parkinson's disease and those which are preserved are described in the next paragraph.

Vulnerability factors of dopaminergic neurons in Parkinson's disease

Various mechanisms may account for the selective vulnerability of some dopaminergic neurons in Parkinson's disease. Several lines of evidence suggest that the metabolism of oxygen free radicals and that of calcium is of particular interest. Indeed, increased intracellular concentrations of these two compounds have been shown to be involved in or to provoke nerve cell death under several different experimental conditions.

A higher production of oxygen free radicals and/or a low level protection against them in the regions most at risk may represent the best hypothesis to explain the role of oxidative stress in the selective vulnerability of some dopaminergic neurons. Indeed, two populations of dopaminergic neurons are observed in the human mesencephalon, those containing neuromelanin and those without. This compound is synthesized by auto-oxidation of dopamine, a reaction which is associated with the production of highly toxic oxygen free radicals. The melanized dopaminergic neurons may thus produce more free radicals than those without melanin. Interestingly, the melanized dopaminergic neurons are found in the dopaminergic cell groups which are the most affected in Parkinson's disease, whereas the non-melanized ones are located in the least affected regions (Hirsch et al., 1988). This suggests that the dopaminergic neurons which produce the higher amount of oxygen free radicals are those which are the most vulnerable. Moreover, the most vulnerable dopaminergic neurons are also those which are poorly protected against oxidative stress. Indeed, all the mesencephalic dopaminergic neurons contain copper/zinc- and manganese-dependent superoxide dismutases, which are enzymes involved in the transformation of the superoxide ion into hydrogen peroxide in the cytoplasm and the mitochondria, respectively (Ceballos et al., 1990; Zhang et al., 1994). Yet, the density of glutathione peroxidase-positive cells is lower in the dopaminergic cell groups which are vulnerable in Parkinson's disease than in the cell groups which are preserved (Damier et al., 1993). Since hydrogen peroxide is mostly catalyzed by glutathione peroxidase in the brain and this compound is the main source of the extremely toxic hydroxyl ion, this suggests that the most vulnerable dopaminergic neurons in Parkinson's disease are particularly sensitivite to oxidative stress.

The dopaminergic neurons which degenerate in Parkinson's disease also seem to be particularly sensitive to a rise in intracellular calcium concentrations. Indeed, calbindin D28K, a protein that binds more than 95% of the intracellular calcium, has been found only in the dopaminergic cell groups that are preserved in Parkinson's disease and in non-human models of Parkinson's disease (Yamada et al., 1990; Manaye et al., 1991; Hirsch et al., 1992). The association between the presence of calbindin in dopaminergic neurons and their preservation in parkinsonian syndromes is even observed within the substantia nigra pars compacta, since the dorsally located neurons, which are more preserved, express this protein whereas the more ventrally located ones, which degenerate, are devoid of the protein. In addition, this relationship seems to be specific to calbindin, since it has not been observed for calretinin, another calcium binding protein (Mouatt-Prigent et al., 1994).

The data concerning the free radical defense mechanisms and regulation of the intracellular calcium homeostasis may explain why some catecholaminergic neurons degenerate in the mesencephalon of patients with Parkinson's disease and others do not. Such a hypothesis does not explain the primary cause of neuronal degeneration in the disease but may provide clues to identifying the cascade of events leading to nerve cell death. Indeed, it suggests that oxidative stress or increased intracellular calcium concentrations may be associated with nerve cell death. The data supporting this supposition are discussed in the next paragraph.

Factors possibly involved in nerve cell death in Parkinson's disease

The exact mechanism of nerve cell death in Parkinson's disease is still unknown. Among several hypotheses, it has been proposed that oxidative stress may participate in the nerve cell death. Indeed, it has long been known that oxygen reactive molecules can damage several biological molecules, including carbohydrates, proteins, nucleic acid and lipids (Fahn and Cohen, 1992). This is supported by the fact that an increased lipid peroxidation, interpreted as the consequence of an attack on polyunsaturated fatty acid by oxygen free radicals, has been reported in the substantia nigra of patients with Parkinson's disease (Dexter et al., 1989). Excessive amounts of oxygen free radicals can be produced when transition metals, especially iron, interact with hydrogen peroxide to form highly reactive hydroxyl ions. Interestingly, several studies have reported an increased iron content in the parkinsonian substantia nigra (Earle, 1968; Dexter et al., 1989; Jellinger et al., 1990, 1992; Hirsch et al., 1991; Good et al., 1992), suggesting a possible role of iron in the mechanism of nerve cell death in Parkinson's disease. This increase seems to be specific to (1) the altered dopaminergic cell groups, since it is not observed in the central gray substance which, is unaffected in Parkinson's disease, and (2) the disease, since it is not detected in the substantia nigra of patients with progressive supranuclear palsy, in which dopaminergic neurons also degenerate, probably through a different mechanism (Hirsch et al., 1991).This metal may be particularly active in inducing oxidative damage in the substantia nigra because, even in normal physiological circumstances, the neurons located in this structure produce large amounts of oxygen free radicals and are already poor metabolizer of free radicals (see above).

The origin of the increased iron concentration in the substantia nigra of patients with Parkinson's disease is not known. One of the possible causes could be a reduced concentration of iron-binding proteins, such as ferritin, or an increased cellular iron uptake. Ferritin levels have been measured in the substantia nigra of patients with Parkinson's disease, but contradictory results showing either reduced or unchanged ferritin levels in the substantia nigra of patients with Parkinson's disease have been reported (Riederer et al., 1989; Dexter et al., 1990). Such differences, which are probably due to the use of antibodies directed against the heavy or light chains of ferritin, call for more extensive studies of neuronal iron-binding capacities in Parkinson's disease. More recently, putative iron uptake mechanisms in this disease have also been analyzed. By analogy to the mechanism of iron penetration in the liver cells, a mechanism mediated by transferrin and its receptor might account for iron penetration into nigral dopaminergic neurons. However, this hypothesis is unlikely given the extremely low density of ^{125}I-transferrin binding sites in the human substantia nigra and their unchanged levels in Parkinson's disease (Faucheux et al., 1993). Alternatively, iron may gain access to dopaminergic neurons at the level of their terminals in the striatum and may be retrogradely transported to their cell bodies in the substantia nigra. Such a hypothesis is compatible with the high density of ^{125}I-transferrin binding sites observed in the striatum and their increased density in Parkinson's disease (Faucheux et al., 1995). Caution is needed in interpreting these results, however, since the increased density of transferrin receptors may occur not only on dopaminergic terminals but also on those of other cell types. Lactoferrin (previously called lactotransferrin) and its receptor, which is specific for this type of transferrin, may also participate in iron penetration into dopaminergic neurons. Indeed, using specific antibodies we found the receptor for lactoferrin to be localized on neurons (pericarya, dendrites and axons), cerebral microvasculature, and, in some cases, glial cells in the human substantia nigra (Fig. 2) (Faucheux et al., 1995). In Parkinson's disease, the immunoreactivity was increased on neurons and microvessels in the substantia nigra and the other affected dopaminergic cell groups but not in the third cranial nerve nucleus, which is unaffected by the disease. Moreover, this increase was more pronounced in the most severely affected patients. These data suggest 1) that lactoferrin receptors may be involved in iron uptake in the human brain both at the level of the blood-brain barrier and on the cytoplasmic membrane of the neurons, 2) that alterations in this iron entry system may account for the increased iron levels in nigral neurons of patients with Parkinson's disease. A role of lactoferrin and its receptor in the succession of biochemical changes leading to nerve cell death is further supported by the fact that lactoferrin levels have also been found to be increased in the substantia nigra of patients with Parkinson's disease (Leveugle et al., 1996).

Besides oxidative stress, other factors are also thought to participate in the mechanism of nerve cell death. This is in particularly true of calcium, which can activate proteases or endonucleases and ultimately lead to cell death. As indicated above, the dopaminergic neurons which degenerate in Parkinson's

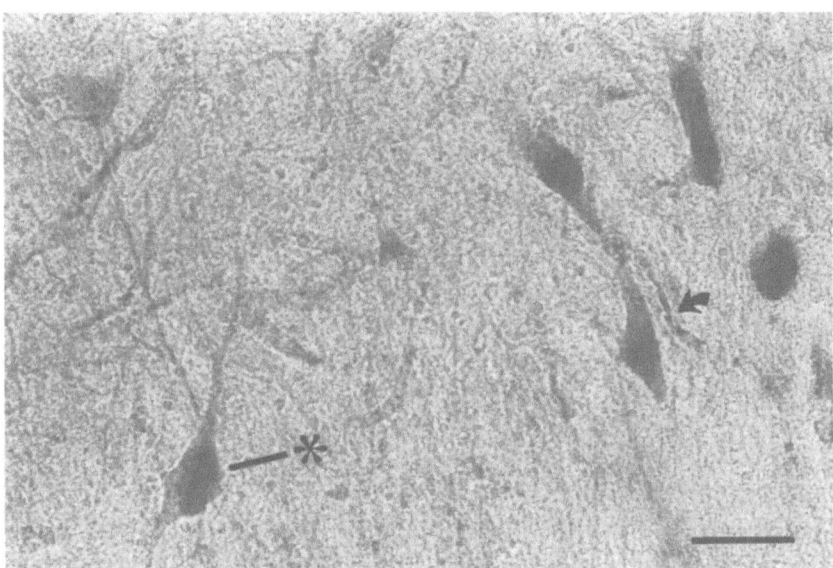

Fig. 2. Lactoferrin receptor immunoreactivity in the substantia nigra of a control subject. Note the presence of immunostained neurons (asterisk) and blood vessels (arrow). Bar indicates 50 µm

disease are particularly sensitive to a rise in intracellular concentrations. Yet, the involvement of calcium in the various events leading to nerve cell death in Parkinson's disease has yet to be proven. We addressed this problem by analysing the distribution and expression of the calcium-dependent protease m-calpain in the mesencephalon of patients with Parkinson's disease and that of control subjects. Two calpain types have been described, one referred to as calpain I or (μ-calpain), which is mainly concentrated in neuronal cell bodies (Sim et al., 1985) and is active at micromolar concentration, and the other, designated calpain II (or m-calpain), which is predominantly observed in axon tracts and glial cells and requires at least millimolar concentration of calcium (Nixon et al., 1986). Quantitative analysis of m-calpain immunoreactivity in parkinsonian patients revealed an increase in the number of m-calpain-positive neurons in the substantia nigra and locus ceruleus, compared to control subjects (Mouatt-Prigent et al., 1996). The density of m-calpain-positive fibers was also increased in both structures. Moreover, calpain-positive fibers presenting an abnormal morphology, and which were also ubiquitin-positive, were observed in the parkinsonian substantia nigra and locus ceruleus, suggesting that the presence of m-calpain is associated with degenerating neurons (Fig. 3). This is further supported by the presence of calpain immunoreactivity in the Lewy bodies, which are the histopathological stigmata of the disease. Furthermore, an unaltered density of m-calpain-positive neurons and nerve fibers was observed in progressive supranuclear palsy and striatonigral degen-

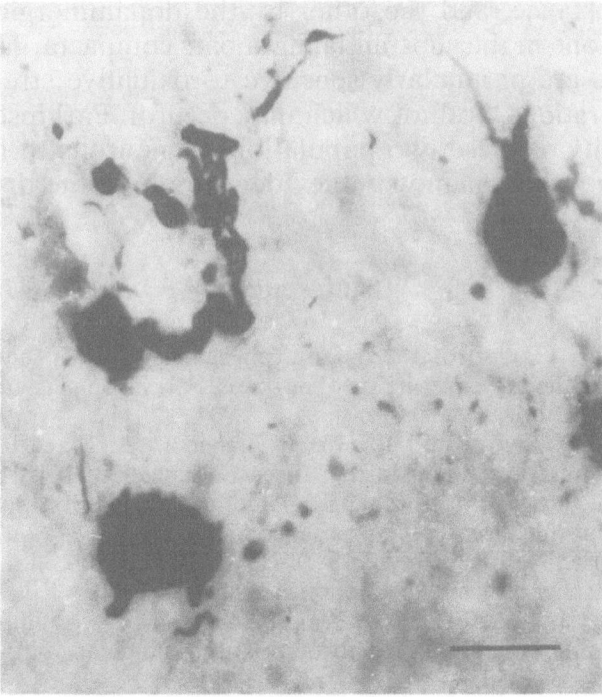

Fig. 3. Calpain immunostained nerve fibers displaying an abnormal morphology in the locus coeruleus of a patients with idiopathic Parkinson's disease. Bar indicates 100 μm

eration, suggesting that the increased calpain expression is specific to a given type of nerve cell death mechanism. However, this supposition must be treated with caution given 1) the loss of neurons expressing calpain in the cerebral cortex of patients with Alzheimer's disease (Iwamoto et al., 1991), 2) the presence of calpain in senile plaques and neurons bearing neurofibrillary tangles in the cerebral cortex of patients with Alzheimer's disease (Iwamoto et al., 1991), and 3) the possible involvement of calpain in nerve cell death observed in experimental ischemia and hypoxia (Nilsson et al. 1990; Lee et al., 1991; Ostwald et al., 1993; Rami et al., 1993; Yoshida et al., 1995). Whatever the case, the increased expression of calpain, probably the reflection of an increased intracellular calcium concentration observed in the catecholaminergic cell groups affected in Parkinson's disease, may participate in the mechanism of nerve cell death. This rise in intracellular calcium concentration may provoke deleterious effects, such as the activation of endonucleases and phospholipases. Nevertheless, whether increased m-calpain concentrations represent an early or a late event in the mechanism leading to nerve cell death, and whether they occur earlier than oxidative stress, remains to be established.

Conclusion

The above data suggest that two populations of dopaminergic neurons may exist in the mesencephalon: one population represented by the neurons of the central gray substance, and the other by the dopaminergic neurons of the calbindin-poor zone in the substantia nigra pars compacta. These two populations of neurons are particularly sensitive to oxidative stress and increased calcium concentrations, both of which may occur in Parkinson's disease. The great vulnerability of these two populations of neurons to these deleterious compounds may well explain why they degenerate in the disease.

References

Bogerts B, Häntsch J, Herzer M (1983) A morphometric study of the dopamine-containing cell groups in the mesencephalon of normals, Parkinson patients, and schizophrenics. Biol Psych 18: 951–969

Ceballos I, Lafon M, Javoy-Agid F, Hirsch E, Nicole A, Sinet PM, Agid Y (1990) Superoxide dismutase and Parkinson's disease. Lancet 335 i: 1035–1036

Damier P, Hirsch E, Javoy-Agid F, Zhang P, Agid Y (1993) Glutathione peroxidase, glial cells and Parkinson's disease. Neuroscience 52: 1–6

Damier P, Hirsch EC, Agid Y, Graybiel AM (1995) Pattern of cell loss in the substantia nigra in Parkinson's disease. Soc Neurosci Abstr 21: 1250

Dexter DT, Carter CJ, Wells FR, Javoy-Agid F, Agid Y, Lees A, Jenner P, Marsden CD (1989) Basal lipid peroxidation in substantia nigra is increased in Parkinson's disease. J Neurochem 52: 381–389

Dexter DT, Wells FR, Lees AJ, Agid F, Agid Y, Jenner P, Marsden CD (1989) Increased nigral iron content and alterations in other metal ions occurring in brain in Parkinson's disease. J Neurochem 52: 1830–1836

Dexter DT, Carayon A, Vidailhet M, Ruberg M, Agid F, Agid Y, Lees AJ, Wells FR, Jenner P, Marsden CD (1990) Decreased ferritin levels in brain in Parkinson's disease. J Neurochem 55: 16–20

Earle KM (1968) Studies in Parkinson's disease including X-ray fluorescent spectroscopy of formalin fixed tissue. J Neuropathol Exp Neurol 27: 1–14

Fahn S, Cohen G (1992) The oxidant stress hypothesis in Parkinson's disease: evidence supporting it. Ann Neurol 32: 796–798

Faucheux BA, Hirsch EC, Villares J, Selimi F, Mouatt-Prigent A, Javoy-Agid F, Agid Y (1993) Distribution of ^{125}I-ferrotransferrin binding sites in the mesencephalon of control subjects and patients with Parkinson's disease. J Neurochem 60: 2238–2241

Faucheux BA, Herrero MT, Villares J, Levy R, Javoy-Agid F, Obeso JA, Hauw JJ, Agid Y, Hirsch EC (1995) Autoradiographic localization and density of [^{125}I]ferrotransferrin binding sites in the basal ganglia of control subjects, patients with Parkinson's disease and MPTP-lesioned monkeys. Brain Res 691: 115–124

Faucheux BA, Nillesse N, Damier P, Spik G, Mouatt-Prigent A, Pierce A, Leveugle B, Kubis N, Hauw JJ, Agid Y, Hirsch EC (1995) Expression of lactoferrin receptors is increased in the mesencephalon of patients with Parkinson's disease. Proc Natl Acad Sci USA 92: 9303–9307

Fearnley JM, Lees AJ (1991) Ageing and Parkinson's disease: substantia nigra regional selectivity. Brain 114: 2283–2301

Hassler R (1937) Zur Normalanatomie der Substantia Nigra, Versuch einer architektonischen Gliederung. J Psychol Neurol 48: 1–55

Hassler R (1938) Zur Pathologies der Paralysis Agitans und des post enzephalitischen Parkinsonismus. J Psychol Neurol 48: 387–476

Hirsch EC, Graybiel AM, Agid Y (1988) Melanized dopaminergic neurons are differentially affected in Parkinson's disease. Nature 334: 345–348

Hirsch EC, Brandel JP, Galle P, Javoy-Agid F, Agid Y (1991) Iron and aluminum increase in the substantia nigra of patients with Parkinson's disease: an x-ray microanalysis. J Neurochem 56: 446–451

Hirsch EC, Mouatt A, Thomasset M, Javoy-Agid F, Agid Y, Graybiel AM (1992) Expression of calbindin D_{28K}-like immunoreactivity in catecholaminergic cell groups of the human midbrain: normal distribution and distribution in Parkinson's disease. Neurodegeneration 1: 83–93

Ito H, Goto S, Sakamoto S, Hirano A (1992) Calbindin-D28K in the basal ganglia of patients with Parkinsonism. Ann Neurol 32: 543–550

Iwamoto N, Thangnipon W, Crawford C, Emson PC (1991) Localization of calpain immunoreactivity in senile plaques and in neurones undergoing neurofibrillary degeneration in Alzheimer's disease. Brain Res 561: 177–180

Jellinger K, Paulus W, Grundke-Iqbal I (1990) Brain iron and ferritin in Parkinson's and Alzheimer's diseases. J Neural Transm 2: 327–340

Jellinger K, Kienzl E, Rumpelmair G, Riederer P, Stachelberger H, Ben-Shachar D, Youdim MB (1992) Iron-melanin complex in substantia nigra of parkinsonian brains: an X-ray microanalysis. J Neurochem 59: 1168–1171

Lee KS, Frank S, Vanderklish P, Arai A, Lynch G (1991) Inhibition of proteolysis protects hippocampal neurons from ischemia. Proc Natl Acad Sci USA 88: 7233–7237

Leveugle B, Faucheux BA, Bouras C, Nillesse N, Spik G, Hirsch EC, Agid Y, Hof PR (1996) Immunohistochemical analysis of the iron binding protein lactotransferrin in the mesencephalon of Parkinson's disease cases. Acta Neuropathol 91: 566–572

Manaye KF, Sonsalla PK, Brooks BA, German DC (1991) Calbindin-$_{28k}$ is located in the midbrain dopaminergic neurons which are resistant to MPTP-induced degeneration. Soc Neurosci Abstr 17: 1275

Mouatt-Prigent A, Agid Y, Hirsch EC (1994) Does the calcium binding protein calretinin protect dopaminergic neurons against degeneration in Parkinson's disease? Brain Res 668: 62–70

Mouatt-Prigent A, Karlsson JO, Agid Y, Hirsch EC (1996) Increased m-calpain expression in the mesencephalon of patients with Parkinson's disease but not in other neurodegenerative disorders involving the mesencephalon: a role in cell death? Neuroscience 73: 979–987

Nilsson E, Alafuzoff I, Blennow K, Blomgren K, Hall C M, Janson I, Karlsson I, Wallin A, Gottfries CG, Karlsson JO (1990) Calpain and calpastatin in normal and Alzheimer-degenerated human brain tissue. Neurobiol Aging 11: 425–431

Nixon RA, Quackenbush R, Vitto A (1986) Multiple calcium-activated neutral proteinases (CANP) in mouse retinal ganglion cell neurons: specificities for endogenous neuronal substrates and comparison to purified brain CANP. J Neurosci 6: 1252–1263

Ostwald K, Hagberg H, Andine P, Karlsson JO (1993) Up-regulation of calpain activity in neonatal rat brain after hypoxic-ischemia. Brain Res 630: 289–294

Rami A, Krieglstein J (1993) Protective effects of calpain inhibitors against neuronal damage caused by cytotoxic hypoxia in vitro and ischemia in vivo. Brain Res 609: 67–70

Riederer P, Sofic E, Rausch WD, Schmidt B, Reynolds GP, Jellinger K, Youdim MBH (1989) Transition metals, ferritin, glutathione, and ascorbic acid in parkinsonian brains. J Neurochem 52: 515–520

Siman R, Gall C, Perlmuter LS, Christian C, Baudry M, Lynch G (1985) Distribution of calpain I, an enzyme associated with degenerative activity, in rat brain. Brain Res 347: 399–403

Yamada T, McGeer PL, Baimbridge KG, McGeer P (1990) Relative sparing in Parkinson's disease of substantia nigra dopamine neurons containing calbindin D28k. Brain Res 26: 303–307

Yoshida KI, Sorimachi Y, Fujiwara M, Hironaka K (1995) Calpain is implicated in rat myocardial injury after ischemia or reperfusion. Jpn Circ J 59: 40–48

Zhang P, Anglade P, Hirsch EC, Javoy-Agid F, Agid Y (1994) Distribution of manganese dependent superoxide dismutase in the human brain. Neuroscience 61: 317–330

Authors' address: Dr. Etienne C. Hirsch, INSERM U289, Hôpital de la Salpêtrière, 47 Boulevard de l'Hôpital, F-75013 Paris, France.

N-Methyl-(R)salsolinol as a dopaminergic neurotoxin: From an animal model to an early marker of Parkinson's disease

M. Naoi[1], **W. Maruyama**[2], **P. Dostert**[3], and **Y. Hashizume**[4]

[1]Department of Biosciences, Nagoya Institute of Technology, Nagoya, Japan
[2]Department of Neurology, Nagoya University School of Medicine, Nagoya, Japan
[3]Research and Development, Pharmacia and Upjohn, Milan, Italy
[4]Institute for Medical Science of Aging, Aichi Medical University, Aichi, Japan

Summary. A dopamine-derived 1(R),2(N)-dimethyl-6,7-dihydroxy-1,2,3,4-tetrahydrosioquinoline [N-methyl-(R)salsolinol] was found to occur enantio-selectively in human brain. This isoquinoline induced parkinsonism in rat after injection in the striatum, and the behavioral, biochemical and pathological changes were very similar to those in Parkinson's disease. N-Methyl-(R)salsolinol depleted dopamine neurons in the rat substantia nigra without necrotic tissue reaction, which may be due to the apoptotic death process, as proved by its induction of DNA damage in dopaminergic neuroblastoma SH-SY5Y cells. N-Methyl-(R)salsolinol was found to increase significantly in the cerebrospinal fluid of parkinsonian patients. All these results suggest that N-methyl-(R)salsolinol may be an endogenous neurotoxin to cause Parkinson's disease and the enzymes involved in its biosynthesis and catabolism may be endogenous factors in the pathogenesis of this disease.

Parkinson's disease is characterized by selective degeneration of dopamine neurons in the substantia nigra, whose pathogenesis still remains to be elucidated. 1-Methyl-4-phenyl-1,2,3,6-tetrahydropyridine (MPTP) is well known to cause parkinsonism in humans with neurotoxicity specific for dopamine neurons (for reviews, see Langston et al., 1987; Singer et al., 1987; Tipton and Singer, 1993). MPTP is oxidized by type B monoamine oxidase [monoamine: oxygen oxidoreductase (deaminating), EC 1.4.3.4] into a potent neurotoxin, 1-methyl-4-phenylpyridinium ion (MPP$^+$) (Chiba et al., 1984). The oxidation into MPP$^+$ is essential for the specificity and potency of the neurotoxicity to dopamine neurons. The results with MPTP suggest that there may be endogenous or xenobiotic neurotoxins to cause Parkinson's disease in humans. As candidates of naturally-occurring MPTP-like compounds, 6,7-dihydroxy-1,2,3,4-tetrahydroisoquinolines (Naoi et al., 1994, 1995a, 1995b) are now the most promising neurotoxin specific for dopamine neurons. (R)-Enantiomer of 1-methyl-6,7-dihydroxy-1,2,3,4-tetrahydroisoquinoline (salsolinol) is synthesized from dopamine endogenously in the brain, whereas other proposed neurotoxin candidates, 1,2,3,4-tetrahydroisoquinolines (Nagatsu and Yoshida, 1988) and β-carbolines (Collins et al., 1992), are derived from food. To prove involvement of salsolinol derivatives to the pathogenesis of Parkinson's disease, an animal model of Parkinson's disease was prepared by infusion of catechol salsolinols into the rat striatum. The neurotoxicity was examined by behavioral, biochemical and patho-histological observation. Only 1(R),2(N)-dimethyl-6,7-dihydroxy-1,2,3,4-tetrahydroisoquinoline [N-methyl-(R)salsolinol] could induce parkinsonism in rats. Further more, in the cerebrospinal fluid (CSF) from parkinsonian patients this N-methyl-(R)salsolinol was found to increase selectively, suggesting that it may be involved in the pathogenesis of Parkinson's disease. The in vivo and in vitro study on mechanism of cytotoxicity suggests that N-methyl-(R)salsolinol induces apoptosis in addition to necrosis, through oxidative stress and energy crisis.

Occurrence and metabolism of catechol isoquinolines in human brain

In human brain, 1-methyl-6,7-dihydroxy-1,2,3,4-tetrahydroisoquinoline (salsolinol) (Sjöquist et al., 1982) and N-methyl-salsolinol (Niwa et al., 1991) have been identified. We recently found that only the (R)-enantiomer of salsolinol and N-methyl-salsolinol can be detected in human brain (Deng et al., 1995). Also in the intraventricular fluid (IVF) and cerebrospinal fluid (CSF), the predominant presence of the (R)-enantiomers was confirmed (Maruyama et al., 1996). As shown in Fig. 1, the enantio-specific biosynthesis of (R)-salsolinol has been previously proposed to occur by condensation of dopamine with pyruvic acid, followed by enzymatic decarboxylation and reduction (Dostert et al., 1990). More recently, an enzyme was identified to condense dopamine with acetaldehyde into (R)-salsolinol in human brain cytosol (Naoi et al.,

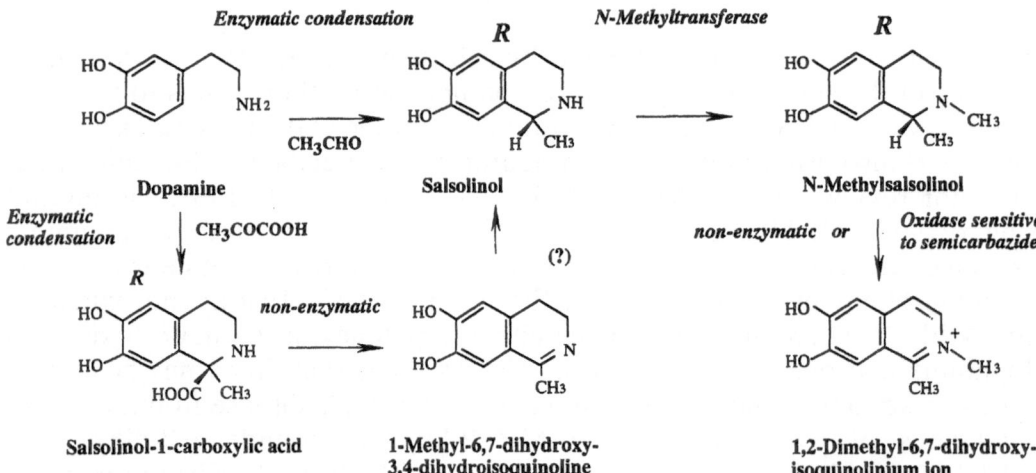

Fig. 1. Biosynthesis of endogenous catechol isoquinolines, (R)salsolinol, N-methyl-(R)-salsolinol and 1,2-dimethyl-6,7-dihydroxyisoquinolinium ion in the brain

1996b), as shown in Fig. 2. The substrate specificity of this enzyme for a biogenic amine, an aldehyde or a keto acid is summarized in Table 1. This enzyme catalyzes the condensation of dopamine with acetaldehyde, formaldehyde or pyruvic acid into (R)-salsolinol, 6,7-dihydroxy-1,2,3,4-tetrahydroisoquinoline (norsalsolinol) or (R)-salsolinol-1-carboxylic acid, whereas N-methyl-dopamine (epinine, deoxyepinephrine) is not a substrate. On the other hand, in wine and other foods (R)- and (S)-salsolinol are identified, which are produced by a non-enzymatic Pictet-Spengler reaction of dopamine with acetaldehyde (Strolin-Benedetti et al., 1989). Even both (R)- and (S)-salsolinol were detected in human plasma (Sällström Baum and Rommelspracher, 1994), salsolinol was reported not to be transported into the brain through the blood-brain barrier (Origitano et al., 1981), which is relevant with our data on the sole presence of the (R)-enantiomers in the brain, IVF and CSF.

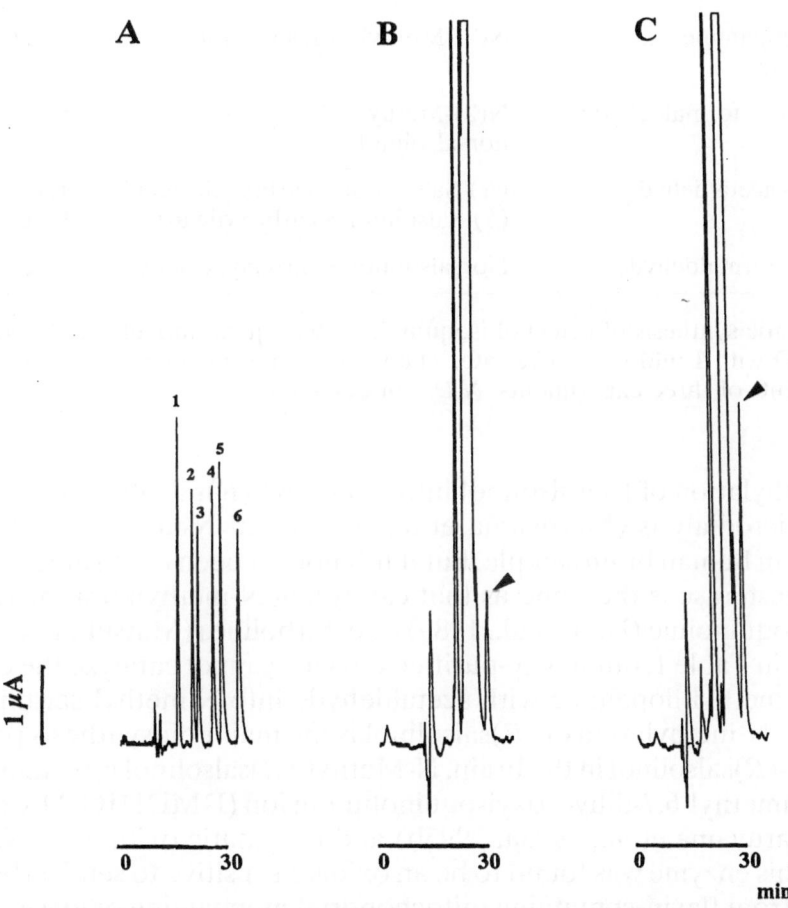

Fig. 2. HPLC patterns of the enzymatic synthesis of (R)salsolinol from dopamine and acetaldehyde. **A** Standard (1) salsolinol-1-carboxylic acid; (2) and (3) (S)- and (R)-salsolinol; (4) and (5) (S)- and (R)-N-methyl-salsolinol and (6) 1-methyl-6,7-dihydroxy-3,4-dihydroisoquinoline. **B** and **C** The reaction product after incubation without (B) and with the enzyme sample (C) (0.19 µg protein) with 1 mM dopamine and 2.5 mM acetaldehyde at 37°C for 20 min

M. Naoi et al.

Table 1. Substrate specificity of a purified enzyme to condense an amine with an aldehyde or a keto acid

Substrate	Product	Enzyme activity (nmol/min/mg protein)
Dopamine + acetaldehyde	(R)-Salsolinol	252.7 ± 99.22
	(S)-Salsolinol	N. D.
Dopamine + pyruvic acid	(R)-Salsolinol-1-carboxylic acid	65.1 ± 22.3
	(S)-Salsolinol-1-carboxylic acid	N. D.
	(R)-Salsolinol	81.4 ± 8.67
	(S)-Salsolinol	N. D.
Dopamine + formaldehyde	Norsalsolinol	15.1 ± 4.62
N-Methyl-dopamine + acetaldehyde	N(2)-Methyl-salsolinol	N. D.
N-Methyl-dopamine + formaldehyde	N(2)-Methyl-norsalsolinol	N. D.
Epinephrine + formaldehyde	N(2)-Methyl-3-hydroxy-norsalsolinol	N. D.
L-DOPA + acetaldehyde	(R)-Salsolinol-3-carboxylic acid	1.46 ± 0.30
	(S)-Salsolinol-3-carboxylic acid	N. D.
L-DOPA + formaldehyde	Norsalsolinol-3-carboxylic acid	N. D.

The enzymatic synthesis of catechol isoquinolines was quantitatively analyzed by use of HPLC-ECD with 1 mM each substrate. The value represents mean and SD of quadrate measurements of three experiments. *N.D.* not detected

N-Methylation of (R)salsolinol into N-methyl-(R)salsolinol was proved by in vivo microdialysis (Maruyama et al., 1992). An N-methyltransferase was identified in human brain sample, but it has not yet been confirmed that this N-methyltransferase is the same as that catalyzing N-methylation of 1,2,3,4-tetrahydroisoquinoline (Naoi et al., 1989) or β-carbolines (Matsubara et a., 1992). As shown in Table 1, our newly-purified enzyme cannot catalyze the condensation of N-methyl-dopamine with acetaldehyde into N-methyl-salsolinol, indicating that N-methylation of (R)salsolinol is the major biosynthesis pathway of N-methyl-(R)salsolinol in the brain. N-Methyl-(R)salsolinol is further oxidized into 1,2-dimethyl-6,7-dihydroxyisoquinolinium ion (DMDHIQ$^+$) by non-enzymatic (Maruyama et al., 1995a, 1995b) and enzymatic oxidation (Naoi et al., 1995c). This enzyme was found to be an oxidase sensitive to semicarbazide and different from flavin-containing mitochondrial monoamine oxidase.

These catechol isoquinolines were examined in human brain and the concentration of N-methyl-(R)salsolinol was found to be significantly higher in the striatum than in other brain regions. These results indicate that (R)-salsolinol and its N-methylated derivatives should be enzymatically synthesized from dopamine in situ in dopamine neurons of the human brain.

Preparation of an animal model of Parkinson's disease

Using these 6,7-dihydroxyisoquinoline derivatives, we tried to set up an animal model of Parkinson's disease in rats (Naoi et al., 1996a,b). The dihydroxyisoquinolines were injected into the striatum of male Wistar rats. For chronic continuous infusion, N-methyl-(R)salsolinol was administrated in the striatum for a week by attachment of a cannula to a mini-osmotic pump (Nitta et al., 1993).

Behavioral changes

After a single injection of N-methyl-(R)salsolinol into the left striatum, rats exhibited postural abnormality with the head and trunk deviating toward the lesions site, lateral extension of the right hind limb and stiffness of the tail elevated above the ground. Some rats showed fine regular twitching of right limbs at rest. During spontaneous activity, the rats showed ipsilateral circulation toward the injection site. Injection of N-methyl-(R)salsolinol into both side of the striatum induced akinesia to rats. A single injection of DMDHIQ$^+$ in the left striatum induced hypokinesia in the right limbs, but no involuntary movement was detected. Other 6,7-dihydroxyisoquinolines, N-methyl-(S)-salsolinol, (R)- and (S)-salsolinol, norsalsolinol and N-methyl-norsalsolinol, did not induce any behavioral changes to rats.

Biochemical analysis of monoamines, their metabolites and isoquinolines

Three day after a single injection the rat was sacrificed and the brain was cut into 2 mm-thick slices by coronal section. The first and second (slice # 1 and # 2) and the seventh slice (slice # 7) from the rostral were confirmed to contain the striatum and the substantia nigra, respectively. Catecholamines, indoleamines and their metabolites and the reduced form of 6,7-dihydroxyisoquinolines were analyzed by use of high-performance liquid chromatography (HPLC) with multi electrochemical detector (ECD) system (CEAS, ESA, Chelmsford, MA) (Maruyama et al., 1992; Naoi et al., 1993). DMDHIQ$^+$ was quantitatively analyzed by HPLC-fluorometric detection (Maruyama et al., 1995b; Naoi et al., 1995c). The activity of tyrosine hydroxylase [tyrosine, tetrahydropteridine: oxygen oxidoreductase (3-hydroxylating); EC 1.4.16.2, TH] was measured by HPLC-ECD (Naoi et al., 1988; Maruyama and Naoi, 1994).

Biochemical analyses of monoamines and their metabolites are summarized in Table 2. In the brain slices # 2 containing a portion of the striatum and # 7 containing the substantia nigra, marked reduction in dopamine was observed after injection of N-methyl-(R)salsolinol and DMDHIQ$^+$. The reduction was more manifest with N-methyl-(R)salsolinol than with DMDHIQ$^+$ in slice # 2 (p < 0.05), and in slice # 7 dopamine was lower than the detection limit, 0.025 pmol/injection. Noradrenaline was reduced in the slice # 2 and # 7 after N-methyl-(R)salsolinol injection. By contrast, serotonin was not reduced after injec-

Table 2. Monoamines and their metabolites in the brain 3 days after injection of N-methyl-(R)salsolinol and DMDHIQ[+]

	Monoamine concentration (nmol/g wet weight)							
	DOPA	DA	DOPAC	HVA	NE	MHPG	5-HT	5-HIAA
Slice #1								
Control	18.9 ± 5.6	30.4 ± 12.7	19.4 ± 6.40	2.38 ± 1.21	6.41 ± 3.39	3.72 ± 4.02	0.71 ± 0.40	2.35 ± 1.20
Rats injected with N-methyl-(R)salsolinol	14.9 ± 3.73	5.43 ± 2.33*	12.1 ± 4.92*	1.62 ± 0.75	3.75 ± 2.33*	1.72 ± 1.21*	0.54 ± 0.30	2.70 ± 0.43
Rats injected with DMDHIQ[+]	16.9 ± 4.66	12.2 ± 5.13*	13.2 ± 2.91*	3.59 ± 0.84	5.64 ± 1.72	4.68 ± 2.24	0.45 ± 0.36	4.29 ± 1.10
Slice #2								
Control	12.2 ± 3.3	20.5 ± 6.75	12.4 ± 3.74	1.49 ± 0.48	5.53 ± 2.67	3.60 ± 1.62	0.42 ± 0.26	1.61 ± 0.58
Rat injected with N-methyl-(R)salsolinol	10.6 ± 4.57	3.14 ± 0.38*	7.03 ± 2.49*	1.22 ± 0.47	1.14 ± 0.91*	1.72 ± 1.21*	0.52 ± 0.31	2.14 ± 0.40
Rats injected with DMDHIQ[+]	13.3 ± 1.96	5.59 ± 4.97*	9.33 ± 1.61*	2.96 ± 1.82*.	4.62 ± 1.93	3.15 ± 1.25	0.36 ± 0.31	3.07 ± 1.05
Slice #7								
Control	14.4 ± 2.48	0.34 ± 0.19	0.51 ± 0.25	0.49 ± 0.20	7.31 ± 4.41	2.98 ± 1.01	0.33 ± 0.17	2.09 ± 1.16
Rats injected with N-methyl-(R)salsolinol	8.82 ± 2.42*	0*	0.73 ± 0.54	0.70 ± 0.13	2.81 ± 1.25*	0.58 ± 0.46*	0.52 ± 0.29	3.11 ± 0.63
Rats injected with DMDHIQ[+]	13.9 ± 3.51	0*	0.67 ± 0.29	0.79 ± 0.21	5.06 ± 2.89	2.40 ± 1.26	0.52 ± 0.42	3.34 ± 0.96

DOPA 3,4-dihydroxyphenylalanine; *DA* dopamine; *DOPAC* 3,4-dihydroxyphenylacetic acid; *HVA* 3-methoxy-4-hydroxyphenylacetic acid (homovanillic acid); *NA* noradrenaline; *MHPG* 3-methoxy-4-hydroxyphenylethylene glycol; *5-HT* 5-hydroxytryptamine (serotonin); *5-HIAA* 5-hydroxyindoleacetic acid; * Difference from control is statistically significant (p<0.05)

tion of N-methyl-(*R*)salsolinol. In other brain regions, no significant change in the contents of dopamine, noradrenaline, serotonin and their metabolites was observed after injection of N-methyl-(*R*)salsolinol or DMDHIQ$^+$.

N-Methyl-(*R*)salsolinol and DMDHIQ$^+$ accumulated in rat brain three days after a single injection were quantitatively analyzed in brain slices. After injection of N-methyl-(*R*)salsolinol, this catechol isoquinoline was detected in slice # 1 to 4. The amount of N-methyl-(*R*)salsolinol in slice # 1 and 2 was 1.03 ± 0.17 and 1.71 ± 0.73 pmol/mg wet weight, whereas that of its oxidation product DMDHIQ$^+$ was larger; 3.86 ± 1.23 and 2.38 ± 2.87 pmol/mg wet weight, respectively. In slice # 6 and 7, definite amounts of DMDHIQ$^+$ were found; 0.45 ± 0.16 and 0.48 ± 0.09 pmol/mg wet weight, respectively. After DMDHIQ$^+$ administration, remarkable amounts of the isoquinolinium ion were detected in slice # 1 and # 2, and in slice # 1 the concentration was the highest; 9.80 ± 3.65 pmol/mg wet weight.

The activity of TH was measured in brain slices of rat injected with N-methyl-(*R*)salsolinol and DMDHIQ$^+$. After a single injection of N-methyl-(*R*)salsolinol, in slice # 1 and # 7 TH activity was significantly reduced; 6.46 ± 1.21 and 0.003 ± 0.01 pmol/min/mg wet weight from control 14.1 ± 0.62 and 3.00 ± 1.42 pmol/min/mg protein, respectively ($p < 0.05$), whereas DMDHIQ$^+$ did not affect TH activity. In other brain regions, no significant reduction of TH activity was detected after injection of N-methyl-(*R*)salsolinol or DMDHIQ$^+$.

Histological study

Three days after a single injection of salsolinols or after 1 week of continuous infusion with a mini-osmotic pump in the left side of the striatum, the rat was sacrificed. The brain was cut into sections, which were stained by the hematoxylin-eosin and Klüver-Barrera method. The striatum of the rats after a single injection of N-methyl-(*R*)salsolinol and DMDHIQ$^+$ is shown in Fig. 3. In the striatum of control rats administered with the vehicle, necrosis was not observed, whereas N-methyl-(*R*)salsolinol administration induced only mild necrosis around the injected site (shown by an arrow) with proliferation of blood vessels and appearance of macrophages (Fig. 3A). DMDHIQ$^+$ administration caused massive necrosis around the injected spot and numerous macrophages were observed (Fig. 3B). Klüver-Barrera staining showed also massive destruction of myelin structure after injection of DMDHIQ$^+$, whereas N-methyl-(*R*)salsolinol did not cause significant change in the structure as compared with control.

After 1 week continuous injection of N-methyl-(*R*)salsolinol into the left striatum, immuno-histochemical examination for dopamine neurons was performed by the streptavidin-biotin complex method with rabbit anti-TH antibody. The density of neurons stained with anti-TH antibody was reduced markedly in the substantia nigra of the treated side, as compared with the control side, as shown in Fig. 4. The number of TH-positive neurons was significantly reduced in the substantia nigra of the injected side (3.09 ± 0.32/ 225 μm^2), in comparison with that in the opposite side (6.69 ± 0.62/225 μm^2)

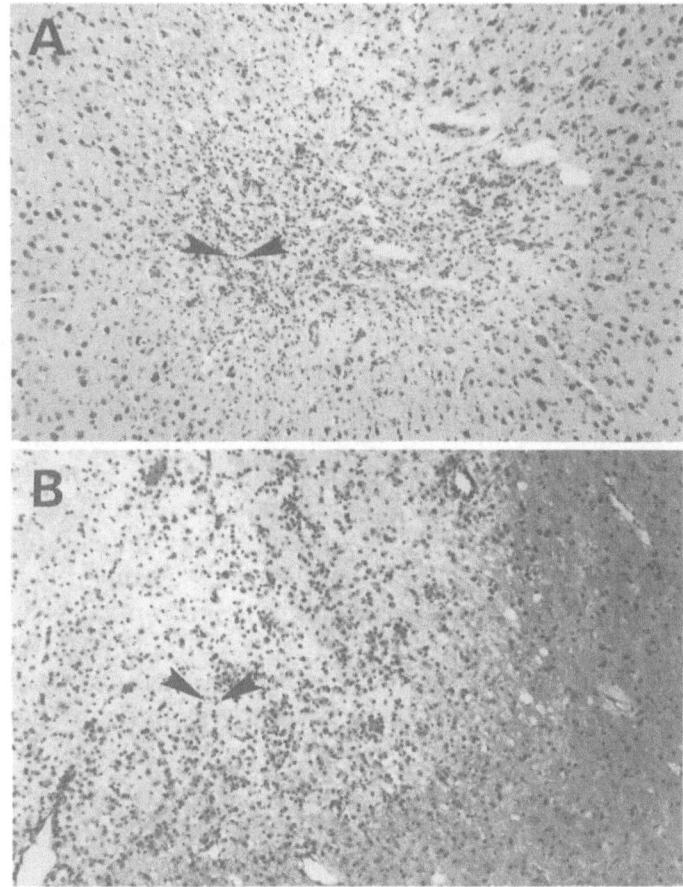

Fig. 3. Morphological changes in the striatum of rats after injection of N-methyl-(*R*)salsolinol and DMDHIQ⁺. Three days after a single injection of the catechol isoquin-oline in the left striatum, the rats were sacrificed and the striatum was stained by hematoxylin-eosin staining. **A** After N-methyl-(*R*)salsolinol injection, the necrosis was limited around the injected site; and **B** DMDHIQ⁺ injection, massive necrosis with activated macrophages. The injected site was shown by an arrow. × 25

(p < 0.05). TH-positive neurons in the ventral segmental area were not affect-ed by the injection of N-methyl-(*R*)salsolinol.

The rat model reported here may be for the first time the setting up of an animal model of Parkinson's disease using an endogenous dihydroxyisoquin-oline present in the human brain. The effects of this compound on the nigro-striatal dopaminergic system were studied systematically by behavioral, bio-chemical and histopathological observations. The results reported here indi-cate that this animal model may be comparable to Parkinson's disease. Previously salsolinol derivatives were examined for the behavioral changes after injection in the brain. Costal et al. (1976) reported that salsolinol and N-methyl-salsolinol induced minor behavioral changes after bilateral injection into the nucleus accumbens of rats. Another dopamine-derived isoquinoline,

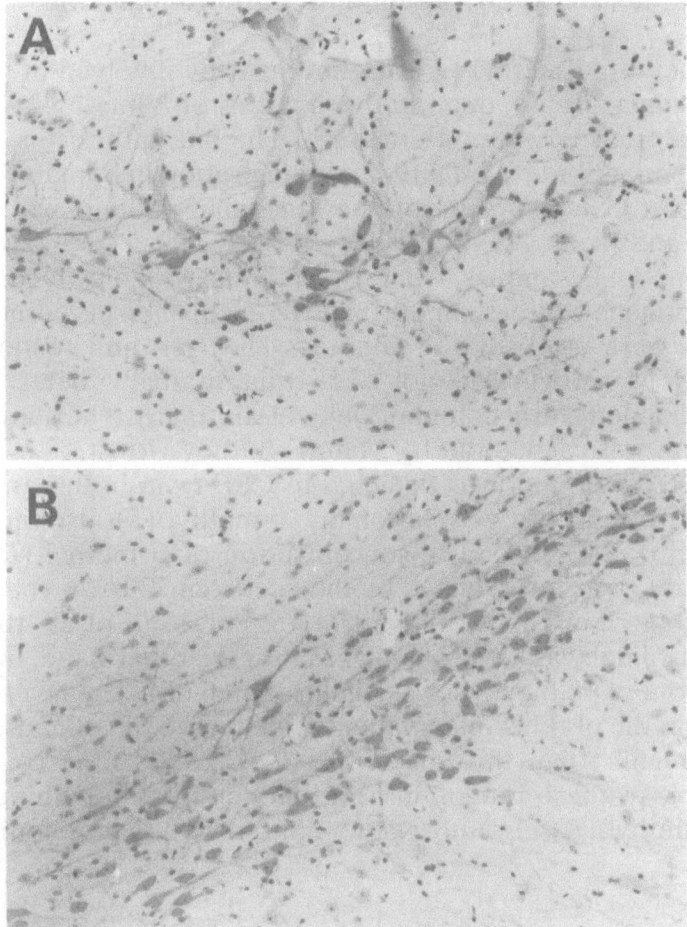

Fig. 4. The substantia nigra of rats injected with N-methyl-(R)salsolinol. Immuno-staining for tyrosine hydroxylase. The rats were sacrificed after 1 week of continuous infusion of N-methyl-(R)salsolinol in the left striatum. **A** The left injected side of the substantia nigra; **B** the right control side. The number of TH positive neurons was decreased in the substantia nigra of the injected side compared with control. × 25

1,2,3,4-tetrahydro-6,7-dihydroxy-1-(3,4-dihydroxybenzyl)-isoquinoline (tetra-hydropapaveroline) was reported to induce behavioral changes to rats after injection in lateral ventricle (Myers et al., 1977). However, no biochemical and pathological data of the brain were present. An adrenaline-derived isoquino-line, 1,2-dimethyl-4,6,7-trihydroxy-1,2,3,4-tetrahydroisoquinoline, was report-ed to induce degeneration of the adrenergic peripheral nerve (Azevedo and Osswald, 1977), and more recently was reported to deplete catecholamines in rat brain after ventricular injection (Liptrot et al., 1993). However, this isoquinoline was synthesized by a Pictet-Spengler reaction and has never been identified in the human brain. Most of these previous studies could not present the selective neurotoxicity of the isoquinolines to dopamine neurons in the nigro-striatal system.

N-Methyl-(*R*)salsolinol in Parkinsonian CSF

To clarify whether N-methyl-(*R*)salsolinol is truly involved in the pathogenesis of Parkinson's disease, quantitative analysis of N-methyl-(*R*)salsolinol in the human materials has been carried out. The biosynthesis of salsolinol was reported to be enhanced by L-DOPA administration in human (Dostert et al., 1989), therefore, CSF samples from parkinsonian patients without any drug therapy were analyzed.

The lumbar CSF samples from 16 patients with newly diagnosed and untreated Parkinson's disease and from 29 control subjects without neurological disorders were used for the analysis. The (*R*)- and (*S*)-enantiomer of salsolinol and N-methyl-salsolinol were quantitatively determined with a cyclodextrin-bonded chiral column (Deng et al., 1995). The statistical evaluation of the data was carried out by Mann-Whitney U test.

N-Methyl-(*R*)salsolinol was detected in CSF from both parkinsonian patients and control. Another enantiomer, N-methyl-(*S*)salsolinol, was under detection limit (< 0.01 nM). The concentration of N-methyl-(*R*)salsolinol in the control group was not affected by the age from 22 to 76 years (r = 0.141). As shown in Fig. 5, in almost all control, except 2 out of 29, the level of N-methyl-(*R*)salsolinol was lower than 6 nM. On the other hand, in 12 parkinsonian patients out of 16 the level was higher than 6 nM. The level of N-methyl-(*R*)salsolinol in the parkinsonian patients (mean and SD, 8.32 ± 2.89 nM) was significantly higher than that in control (4.53 ± 2.08 nM) (p < 0.0001). To estimate the biosynthesis rate of this isoquinoline from dopamine, the concentration of N-methyl-(*R*)salsolinol was compared with that of homovanillic acid

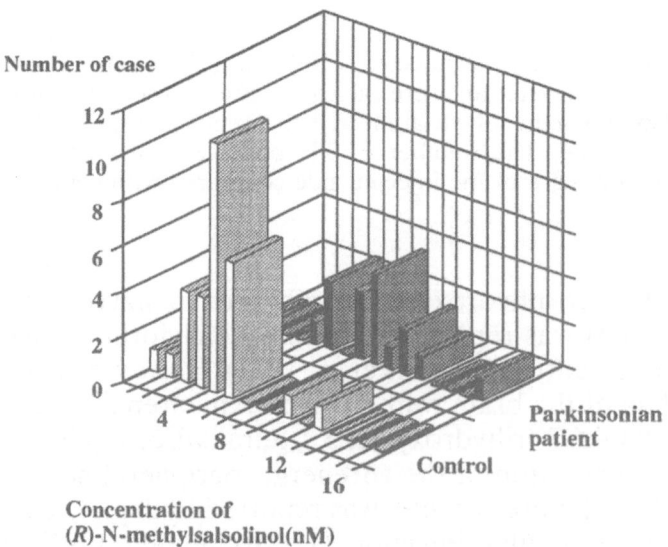

Fig. 5. Histograms of the concentration of N-methyl-(*R*)salsolinol in CSF from parkinsonian patients and control. The concentration of N-methyl-(*R*)Salsolinol was expressed as nM, and the number of patients with the corresponding concentration was shown in columns

(HVA), a major metabolite of dopamine. The ratio also significantly increased in parkinsonian patients; 0.086 ± 0.067 from 0.034 ± 0.030 in control ($p < 0.0002$). In control group, N-methyl-(R)salsolinol concentration tended to increase according to HVA level, and in parkinsonian patients the correlation between the value of N-methyl-(R)salsolinol and that of HVA was not observed.

After L-DOPA administration the same amounts of (R)- and (S)-salsolinol were detected in human urine (Dostert et al., 1989), but our data proved the predominant presence of (R)-enantiomer of salsolinol and N-methyl-salsolinol in the brain, indicating again that (S)-enantiomers could not be transported into the brain. Involvement of the enzymes in the biosynthesis of (R)-enantiomers is also suggested by the fact that in the parkinsonian brain level of dopamine, a precursor of salsolinol, is significantly reduced. The increase in N-methyl-(R)salsolinol levels in CSF and probably in the brain may result from a specific biochemical process, that is, its increased synthesis and/or reduced catabolism, as supported by the increase in the ratio of N-methyl-(R)salsolinol to HVA in parkinsonian CSF. As described above, two enzymes arc now considered to be involved in the synthesis of N-methyl-(R)salsolinol; an enzyme catalyzing the condensation of dopamine with acetaldehyde or pyruvic acid to yield (R)salsolinol [Naoi et al., 1996b], and an N-methyltransferase which catalyzes N-methylation of (R)-salsolinol (Maruyama et al., 1992).

Sustained high level of N-methyl-(R)salsolinol in the brain might induce degeneration of the neurons after a long term. In addition, analysis of this dopaminergic neurotoxin in CSF might be applicable as a pre-clinical early marker for Parkinson's disease. The study to clarify the biological factors which determine the level of this neurotoxin in each individual is now under the way.

Mechanism of selective neurotoxicity of N-methyl-(R)salsolinol

Histological examination in the striatum indicates that DMDHIQ$^+$ was much more cytotoxic than its reduced precursor N-methyl-(R)salsolinol: massive necrosis and destruction of myelin structure were observed, while N-methyl-(R)salsolinol caused mild extent of necrosis in a limited area. However, the cytotoxicity of DMDHIQ$^+$ was not selective to dopaminergic neurons as shown by morphological observation and biochemical analysis. In the striatum, N-methyl-(R)salsolinol depleted dopamine and noradrenaline more markedly than DMDHIQ$^+$ and TH activity was reduced by injection of N-methyl-(R)salsolinol, but not DMDHIQ$^+$. After injection of N-methyl-(R)salsolinol in the striatum, DMDHIQ$^+$ was found to accumulate in the substantia nigra in addition to the striatum. These results seem to be contradictory, but may be elucidated as follows. Selective uptake of N-methyl-(R)salsolinol determines the selectivity, and its oxidation produces hydroxyl radical and potent cytotoxic DMDHIQ$^+$, which can accumulate in the neurons more markedly than N-methyl-(R)salsolinol.

Selective uptake of N-methyl-(R)salsolinol was confirmed using human dopaminergic neuroblastoma SH-SY5Y cells. Only N-methyl-(R)salsolinol was found to be taken up in the cells by a dopamine transport system, whereas N-methyl-(S)salsolinol and other 6,7-dihydroxyisoquinolines were not (Takahashi et al., 1994). N-Methyl-(R)salsolinol was found to generate hydroxyl radical during its auto-oxidation, as shown by in vivo (Maruyama et al., 1995a) and in vitro experiments (Maruyama et al., 1995b). Fig. 6 shows that non-enzymatic oxidation of N-methyl-(R)salsolinol produces hydroxyl radical and DMDHIQ$^+$, simultaneously. The oxidation product, the positively-charged 6,7-dihydroxyisoquinolinium ion, has been shown to be the most potent cytotoxic to SH-SY5Y cells among catechol isoquinolines examined, as shown in Fig. 7. By use of Alamar Blue assay with a reduction-oxidation indicator, the IC$_{50}$ value of DMDHIQ$^+$ was obtained to be 63 μM, while other isoquinolines had IC$_{50}$ values larger than 500 μM (Takahashi et al., in preparation). In addition, DMDHIQ$^+$ was found to accumulate in cells more markedly than N-methyl-(R)salsolinol by binding to mitochondria or other subcellular compartments (Naoi et al., 1994b). The potent cytotoxicity of DMDHIQ$^+$ causing necrosis in the injected region may be due to the inhibition of the mitochondrial respiratory chain (McNaught et al., 1995) in a similar way as with MPP$^+$ (Mizuno et al., 1987) or N-methylisoquinolinium ion (Suzuki et al., 1992). In human brain DMDHIQ$^+$ may accumulate in the substantia nigra by binding to melanin. As shown by in vitro experiments (Naoi et al., 1994a), DMDHIQ$^+$ was found to bind with melanin with the similar affinity as MPP$^+$ (D'Amato et al., 1987). In addition, the binding of DMDHIQ$^+$ to melanin is regulated by Fe (II) and Fe (III); Fe (II) enhances the binding, whereas Fe (III) releases the

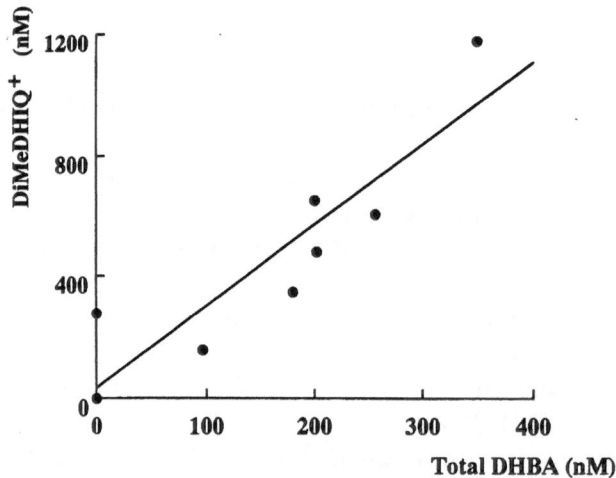

Fig. 6. Simultaneous production of hydroxyl radical and DMDHIQ$^+$ by oxidation of N-methyl-(R)salsolinol. N-Methyl-(R)salsolinol (0.2 to 0.8 mM) was incubated in 100 mM phosphate buffer, pH 7.25, containing 8 mM salicylic acid for 20 min at 37°C. Hydroxyl radical produced was trapped as 2,3- or 2,5-dihydroxybenzoic acid (DHBA) and quantitatively determined by HPLC-ECD. DMDHIQ$^+$ was measured by HPLC-fluorometric detection

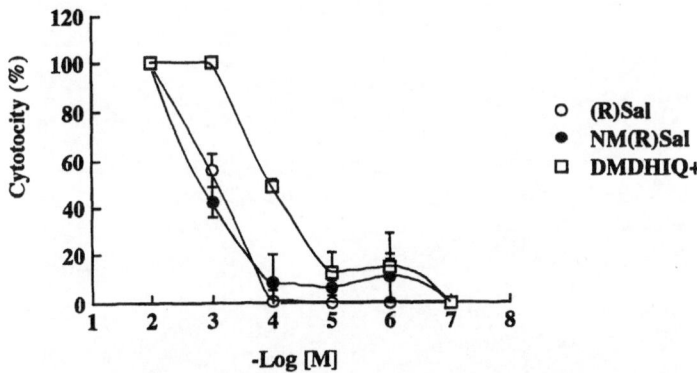

Fig. 7. The cytotoxic effects of dihydroxyisoquinolines to SH-SY5Y cells. The cells were pre-incubated with 10 mM to 0.1 μM isoquinolines for 12 hours. The viability of cells was evaluated by measurement of increase in the specific absorbance due to reduction of an Alamar Blue dye. The cytotoxicity was calculated from the results by comparison of control cells without incubation of isoquinolines

ion from melanin. Fe (III) content in dopamine cells increases in the brain with aging, and also in the brain of patients with Parkinson's disease (Riederer et al., 1989). Age-dependent increase in Fe (III) would cause DMDHIQ$^+$ release from melanin with inhibition of the mitochondrial enzymes and depletion of ATP from dopamine cells as a result.

DNA damage by N-methyl-(R)salsolinol

Apoptotic cell death is now proposed as an important process of neuronal death in addition to necrosis. The selective neurotoxicity of N-methyl-(R)salsolinol was demonstrated further by its induction of DNA damage. SH-SY5Y cells were incubated with catechol isoquinolines and assessed for DNA damage using the single cell gel electrophoresis assay (comet assay) (Ostling and Johanson, 1984; Singh et al., 1988). As shown in Fig. 8, only after incubation of N-methyl-(R)salsolinol, DNA fragmentation was detected. Other catechol isoquinolines, (R)-salsolinol and DMDHIQ$^+$ did not induce DNA damage. The involvement of apoptotic cell death may be relevant with the histopathological observation that no significant necrotic changes could be detected in the substantia nigra, where dopamine neurons were markedly reduced after chronic infusion of N-methyl-(R)salsolinol.

The oxidative stress and energy crisis caused by oxidation of N-methyl-(R)salsolinol may induce necrosis of dopamine neurons in the substantia nigra, as shown here with a rat model of Parkinson's disease. In Parkinson's disease the death of dopaminergic neurons in the substantia nigra is considered to be caused by increased free radical production (Jenner et al., 1992). Increase in Fe (III) and reduction in reduced glutathione levels in the substantia nigra of parkinsonian brain (Riederer et al., 1989) may also enhance the oxidative stress. The studies with MPTP proved that MPP$^+$ inhibits complex I

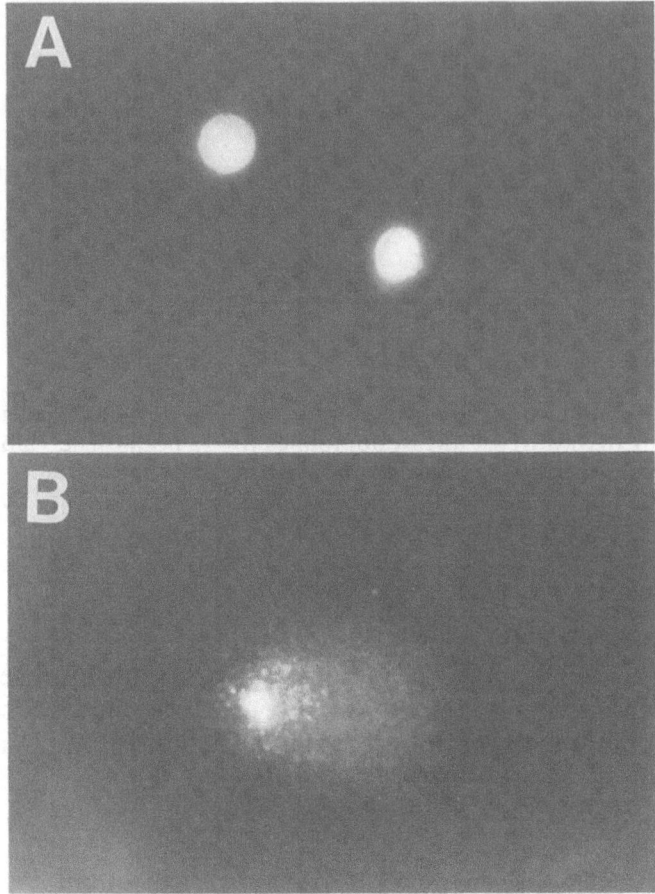

Fig. 8. DNA fragmentation after incubation of SH-SY5Y cells with N-methyl-(*R*)salsoli-
nol. The cells were incubated with 1 mM N-methyl-(*R*)salsolinol and applied to a single-
cell gel electrophoresis. **A** Control and **B** the cells treated with the isoquinoline. DNA
was stained with 4′,6-diamidino-2-phenylindole dihydrochloride, and the fluorescence was
measured

in the mitochondrial respiratory chain and thus induces cell death (Mizuno et
al., 1987). Actually in the parkinsonian brain the deficiency of complex I has
been confirmed (Mizuno et al., 1989; Shapira et al., 1990).

Recently the apoptotic degeneration has been detected in brains of Parkinson's disease (Anglade et al., 1995; Mochizuki et al., 1996). Our data clearly
demonstrate that DNA damage is also induced by this endogenous neurotoxin
N-methyl-(*R*)salsolinol. Hartley et al. (1994) reported that inhibition of complex I by MPP⁺ and rotenone induced apoptosis by a dose-dependent way.
Apoptosis was caused by MPP⁺ at the lower concentrations, whereas necrosis
at the higher concentrations. Our preliminary data also suggest that oxidative
stress may be primarily elicited, followed by DNA fragmentation. These
results indicate that neuron may die not by a single mechanism, but by a
malignant cycle of oxidative stress, energy crisis and apoptosis interacting with
each other.

Our results reported here will not exclude the possible involvement of other endogenous or xenobiotic neurotoxins as the pathogenic agents in human, however at present N-methyl-(*R*)salsolinol seems to satisfy the conditions of an endogenous neurotoxin to induce selective degeneration of dopamine neurons in the substantia nigra of Parkinson's disease.

Acknowledgement

This work was supported by a Grant-In-Aid for Scientific Research on Priority Area from the Ministry of Education, Science and Culture, Japan.

References

Anglade P, Vyas S, Javoy-Agid F, Herrero MT, Michel PP, Marquerz J, Mouatt-Prigient A, Ruberg M, Hirsch EC, Agid Y (1995) Apoptotic degeneration of nigral dopaminergic neurons in Parkinson's disease. Soc Neurosci Abstract 21: 1250

Azevedo I, Osswald W (1977) Adrenergic nerve degeneration induced by condensation products of adrenaline and acetaldehyde. Naunyn-Schmiedeberg's Arch Pharmacol 300: 139–144

Chiba K, Trevor AJ, Castagnoli N Jr (1984) Metabolism of the neurotoxic amine, MPTP, by brain monoamine oxidase. Biochem Biophys Res Commun 120: 574–578

Collins MA, Neafsey EJ, Matsubara K, Cobuzzi R Jr, Rollema H (1992) Indole-N-methylated β-carbolinium ions as potential brain-bioactivated neurotoxins. Brain Res 570: 154–160

Costal B, Naylor RJ, Pinder RM (1976) Hyperactivity induced by tetrahydroisoquinoline derivatives injected into the nucleus accumbens. Eur J Pharmacol 39: 153–160

D'Amato RJ, Alexander GM, Schwartzman RJ, Kitt CA, Price DL, Snyder SH (1987) Evidence for neuromelanin involvement in MPTP-induced neurotoxicity. Nature 327: 324–326

Deng Y, Maruyama W, Dostert P, Takahashi T, Kawai M, Naoi M (1995) Determination of the (*R*)- and (*S*)-enantiomers of salsolinol and N-methylsalsolinol by use of a chiral HPLC column. J Chromatogr B 670: 47–54

Dostert P, Strolin Benedetti M, Dordain G, Vernay D (1989) Enantiometric composition of urinary salsolinol in Parkinsonian patients after Madopar. J Neural Transm [P-D Sect] 1: 269–278

Dostert P, Strolin Benedetti M, Bellotti V, Allievi C, Dordain G (1990) Biosynthesis of salsolinol, a tetrahydroisoquinoline alkaloid, in healthy subjects. J Neural Transm [GenSect] 81: 215–223

Hartley A, Stone JM, Heron C, Cooper JM, Schapira AHV (1994) Complex I inhibitors induce dose-dependent apoptosis in PC12 cells: Relevance to Parkinson's disease. J Neurochem 63: 1987–1990

Jenner P, Schapira AHV, Marsden CD (1992) New insights into the cause of Parkinson's disease. Neurology 42: 2241–2250

Langston JW, Irwin I, Ricaurte GA (1987) Neurotoxins, parkinsonism and Parkinson's disease. Pharmac Ther 32: 19–49

Liptrot J, Holdup D, Phillipson O (1993) 1,2,3,4-Tetrahydro-2-methyl-4,6,7-isoquinolinetriol depletes catecholamines in rat brain. J Neurochem 61: 2199–2206

Maruyama W, Nakahara D, Ota M, Takahashi T, Takahashi A, Nagatsu T, Naoi M (1992) N-Methylation of dopamine-derived 6,7-dihydroxy-1,2,3,4-tetrahydroisoquinoline, (*R*)-salsolinol, in rat brains: in vivo microdialysis study. J Neurochem 59: 395–400

Maruyama W, Naoi M (1994) Inhibition of tyrosine hydroxylase by a dopamine neurotoxin, 1-methyl-4-phenylpyridinium ion: Depletion of allostery to the biopterin cofactor. Life Sci 55: 207–212

Maruyama W, Dostert P, Matsubara K, Naoi M (1995a) N-Methyl(R)salsolinol produces hydroxyl radicals: Involvement to neurotoxicity. Free Rad Biol Med 19: 67–75

Maruyama W, Dostert P, Naoi M (1995b) Dopamine-derived 1-methyl-6,7-dihydroxyisoquinolines as hydroxyl radical promoters and scavengers in the rat brain: In vivo and in vitro studies . J Neurochem 64: 2635–2643

Maruyama W, Abe T, Tohgi H, Dostert P, Naoi M (1996) A dopaminergic neurotoxin, (R)-N-methylsalsolinol, increases in parkinsonian CSF. Ann Neurol 40: 119–122

Matsubara K, Neafsey EJ, Collins MA (1992) Novel S-adenosylmethionine-dependent indole-N-methylation of β-carbolines in brain particulate fractions. J Neurochem 59: 511–518

McNaught K, Thull U, Carrupt P-A, Altomare C, Cellamare S, Carotti A, Testa B, Kenner P, Marsden CD (1995) Inhibition of complex I by isoquinoline derivatives structurally related to 1-methyl-4-phenyl-1,2,3,6-tetrahydropyridine (MPTP). Biochem Pharmacol 50: 1903–1911

Mizuno Y, Saitoh T, Sone N (1987) Inhibition of mitochondrial NADH-ubiquinone oxido-reductase activity by 1-methyl-4-phenyl-pyridinium ion. Biochem Biophys Res Commun 143: 294–299

Mizuno Y, Ohta S, Tanaka M, Takamiya S, Suzuki K, Sato T, Oya H, Ozawa T, Kagawa Y (1989) Deficiencies in complex I subunits of the respiratory chain in Parkinson's disease. Biochem Biophys Res Commun 163: 1450–1455

Mochizuki H, Goto G, Mori H, Mizuno Y (1996) Histochemical detection of apoptosis in Parkinson's disease. J Neurol Sci 137: 120–123

Myers RD, Oblinger MM (1977) Alcohol drinking in the rat induced by acute intracerebral infusion of two tetrahydroisoquinolines and a β-carboline. Drug Alcohol Depend 2: 469–483

Nagatsu T, Yoshida M (1988) An endogenous substance of the brain, tetrahydroisoquinoline, produces parkinsonism in primates with decreased dopamine, tyrosine hydroxylase and biopterin in the nigrostriatal regions. Neurosci Lett 87: 178–182

Naoi M, Takahashi T, Nagatsu T (1988) Simple assay procedure for tyrosine hydroxylase activity by high-performance liquid chromatography employing coulometric detection with minimal sample preparation. J Chromatogr 427: 229–238

Naoi M, Matsuura S, Takahashi T, Nagatsu T (1989) An N-methyltransferase in human brain catalyzes N-methylation of 1,2,3,4-tetrahydroisoquinoline, a precursor of a dopaminergic neurotoxin, N-methylisoquinolinium ion. Biochem Biophys Res Commun 161: 1213–1219

Naoi M, Maruyama W, Acworth IN, Nakahara D, Parvez H (1993) Multi-electrode detection system for determination of neurotransmitters. In: Parvez H, Naoi M, Nagatsu T, Parvez S (eds) Methods in neurotransmitter and neuropeptide research vol I. Elsevier, Amsterdam, pp 1–39

Naoi M, Maruyama W, Dostert P (1994a) Binding of 1,2(N)-dimethyl-6,7-dihydroxyisoquinolinium ion to melanin: effects of ferrous and ferric ion on the binding. Neurosci Lett 171: 9–12

Naoi M, Maruyama W, Niwa T, Nagatsu T (1994b) Novel toxins and Parkinson's disease: N-Methylation and oxidation as metabolic bioactivation of neurotoxin. J Neural Transm [Suppl] 41: 197–205

Naoi M, Maruyama W, Dostert P, Nakahara D, Takahashi T, Nagatsu T (1995a) Metabolic bioactivation of endogenous isoquinolines as dopaminergic neurotoxins to elicit Parkinson's disease. In: Hanin IS, Yoshida M, Fisher A (eds) Alzheimer's and Parkinson's diseases. Recent Developments. Plenum, New York, pp 553–559

Naoi M, Maruyama W and Dostert P (1995b) Dopamine-derived 6,7-dihydroxy-1,2,3,4-tetrahydroisoquinolines: Oxidation and neurotoxicity. Prog Brain Res 106: 227–239

Naoi M, Maruyama W, Zhang JH, Takahashi T, Deng Y, Dostert P (1995c) Enzymatic oxidation of the dopaminergic neurotoxin, 1(R), 2(N)-dimethyl-6,7-dihydroxy-1,2,3,4-tetrahydroisoquinoline, into 1,2(N)-dimethyl-6,7-dihydroxyisoquinolinium ion. Life Sci 57: 1061–1066

Naoi M, Maruyama W, Dostert P, Hashizume Y, Takahashi T, Ota M (1996a) Dopamine-derived endogenous 1(R), 2(N)-dimethyl-6,7-dihydroxy-1,2,3,4- tetrahydroisoquinoline, N-methyl-(R)-salsolinol, induced parkinsonism in rats: Biochemical, pathological and behavioral studies. Brain Res 709: 285–295

Naoi M, Maruyama W, Dostert P, Hashizume Y (1996b) Animal model of Parkinson's disease induced by naturally-occurring 1(R), 2(N)-dimethyl-6,7-dihydroxy-1,2,3,4-tetrahydroisoquinoline. Biogenic Amines 12: 135–147

Nitta A, Murase Y, Furukawa Y, Hayashi K, Hasegawa T, Nabeshima T (1993) Memory impairment and neural dysfunction after continuous infusion of anti-nerve growth factor antibody into the septum in adult rats. Neurosci 57: 495–499

Niwa T, Tekeda T, Yoshizumi H, Tatematsu A, Yoshida M, Dostert P, Naoi M, Nagatsu T (1991) Presence of 2-methyl-6,7-dihydroxy-1,2,3,4-tetrahydroisoquinoline and 1,2-dimethyl-6,7-dihydroxy-1,2,3,4-tetrahydroisoquinoline, novel endogenous amines, in parkinsonian and normal human brains. Biochem Biophys Res Commun 177: 603–609

Origitano T, Hanningen J, Collins MA (1981) Rat brain salsolinol and blood-brain barrier. Brain Res 224: 446–451

Ostling O, Johanson KJ (1984) Microelectrophoretic study of radiation-induced DNA damages in individual mammalian cells. Biochem Biophys Res Commun 123: 291–298

Riederer P, Sofic E, Rausch W-D, Schmidt B, Reynolds GD, Jellinger K, Youdim MBH (1989) Transition metals, ferritin, glutathione, and ascorbic acid in parkinsonian brains. J Neurochem 52: 515–520

Sällström Baum S, Rommelspracher H (1994) Determination of total dopamine, R- and S-salsolinol in human plasma by cyclodextrin bonded-phase liquid chromatography with electrochemical detection. J Chromatogr B 660: 235–241

Schapira AHV, Cooper JM, Dexter D, Jenner P, Clark JB, Marsden CD (1990) Mitochondrial complex I deficiency in Parkinson's disease. J Neurochem 54: 823–827

Singer TP, Castagnoli N Jr, Ramsay RR, Trevor AJ (1987) Biochemical events in the development of parkinsonism induced by 1-methyl-4-phenyl-1,2,3,6-tetrahydropyridine. J Neurochem 49: 1–8

Singh NP, McCoy MT, Tice RR, Schneider EL (1988) A simple technique for quantitation of low levels of DNA damage in individual cells. Exp Cell Res 175: 184–191

Sjöquist B, Eriksson A, Winblad B (1982) Salsolinol and catecholamines in human brain and their relation to alcoholism. Prog Clin Biol Res 90: 57–67

Strolin Benedetti M, Bellotti V, Pianezola E, Moro E, Carminati P, Dostert P (1989) Ratio of the R and S enantiomers of salsolinol in food and human urine. J Neural Transm 77: 47–53

Suzuki K, Mizuno Y, Yamauchi Y, Nagatsu T, Yoshida M (1992) Selective inhibition of complex I by N-methylisoquinolinium ion and N-methyl-1,2,3,4-tetrahydroisoquinoline in isolated mitochondria prepared from mouse brain. J Neurol Sci 109: 219–223

Takahashi T, Deng Y, Maruyama W, Dostert P, Kawai M, Naoi M (1994) Uptake of a neurotoxin-candidate, (R)-1,2-dimethyl-6,7-dihydroxy-1,2,3,4-tetrahydroisoquinoline into human dopaminergic neuroblastoma SH-SY5Y cells by dopamine transport system. J Neural Transm [GenSect] 98: 107–118

Tipton KF, Singer TP (1993) Advances in our understanding of the mechanism of the neurotoxicity of MPTP and related compounds. J Neurochem 61: 1191–1206

Authors' address: Dr. M. Naoi, Department of Biosciences, Nagoya Institute of Technology, Gokiso-cho, Showa-ku, Nagoya 466, Japan.

The halogenated tetrahydro-β-carboline "TaClo":
A progressively-acting neurotoxin

Christine Heim and **K.-H. Sontag**

Department of Psychiatry, University of Göttingen,
Göttingen, Federal Republic of Germany

Summary. "TaClo", (1-trichloromethyl-1,2,3,4-tetrahydro-β-carboline) which is structurally very similar to MPTP (1-methyl-4-phenyl-1,2,3,6-tetrahydropyridine), increases the sensitivity to apomorphine in aging rats after subchronic daily application (0.2 mg/kg i.p.) for 7 weeks. Nine weeks after the last treatment with TaClo, the running speed of the animals during spontaneous nocturnal activity was significantly diminished, and 9 months after the last injection rats developed a pronounced hypersensitivity to apomorphine (0.4 mg/kg s.c.). TaClo, which can be readily produced from endogenous tryptamine and the non-natural aldehyde chloral, appears to be a drug that is able to induce a slowly-developing neurodegenerative process.

Introduction

"TaClo", a new synthetic β-carboline (1-trichloromethyl-1,2,3,4-tetrahydro-β-carboline) (Bringmann et al., 1995) is structurally very similar to MPTP (1-methyl-4-phenyl-1,2,3,6-tetrahydropyridine) and appears to be a drug that is able to induce a slowly-developing neurodegenerative process. The substance is readily produced under quasi-physiological conditions (aqueous solution, pH 7.4, 37°C) from endogenous tryptamine and the non-natural aldehyde chloral. Administration of the drug chloral hydrate, or inhalation of

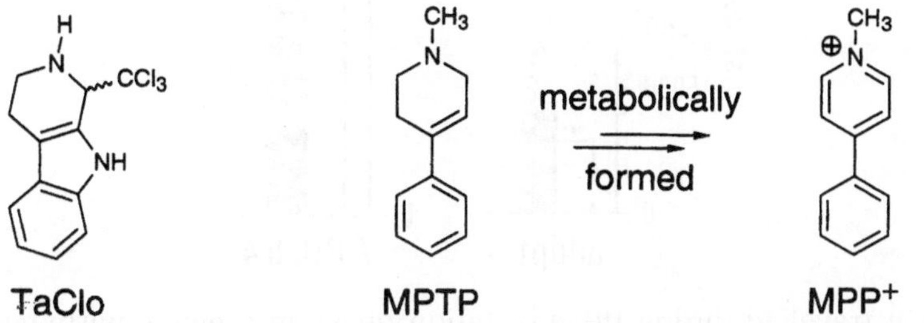

Fig. 1. TaClo (1-trichloromethyl-1,2,3,4-tetrahydro-β-carboline), which shows a great structural similarity to MPTP, appears to be a neurotoxic drug capable of inducing a slowly-developing degenerative process in the dopaminergic system

the industrial solvent trichloroethylene which is metabolised to chloral (Bruckner et al., 1989), can cause spontaneous in vivo formation of TaClo (Bringmann et al., 1995).

The neurotoxic potency of TaClo appears to be different from that of MPTP or its metabolite MPP+. MPTP produces symptoms in monkeys that very closely resemble those of humans suffering from Parkinson's disease (PD), but severe symptoms develop very rapidly after administration of the toxic drug. The administration of MPTP does not therefore appear valid as a model for studying the various stages of the slow progressive illness of PD in man (see also Mohanakumar et al., 1994)

For this reason it is desirable to develop animal models that mimic the slow progressive advance of PD with its late onset of the first typical symptoms. With this in view we have tested the consequences of a subchronic daily application of TaClo on the behaviour of rats and on their reaction in response to the administration of apomorphine for up to several months later.

Results

Figure 2 shows the distance travelled by the rats during the adaptive phase in a new environment 4–9 days after an injection period of 7 weeks in which they each received daily injections of 0.2 mg/kg of TaClo intraperitoneally (i.p.). Rats treated with TaClo exhibited hyperactive locomotion during the exploration and habituation phases within 1 hour, compared to saline-treated controls, but less distance travelled following a subcutaneous application of apomorphine at a dose of 0.4 mg/kg.

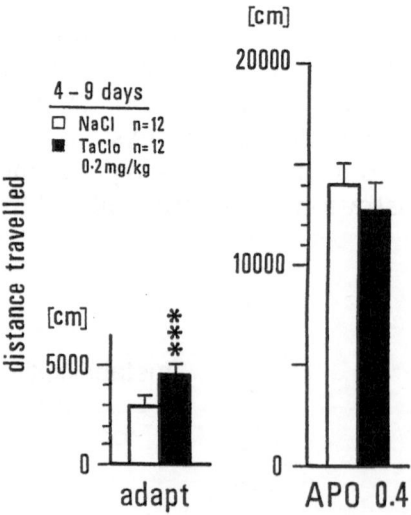

Fig. 2. Distance travelled during the adaptation phase in a new environment 4–9 days after an injection period of seven weeks with a daily injection of 0.2 mg/kg of TaClo i.p. (left), and after 0.4 mg/kg of apomorphine s.c. (right). Statistics: group differences ADAP:$F_{(1,22)}$ = 14.55, p = 0.0002; ANOVA

The running speed during spontaneous nocturnal activity of rats treated with TaClo was significantly diminished 9 weeks after the injection period (not shown). Nine months after TaClo treatment, rats developed a pronounced hypersensitivity to apomorphine as demonstrated by their running behaviour (Fig. 3).

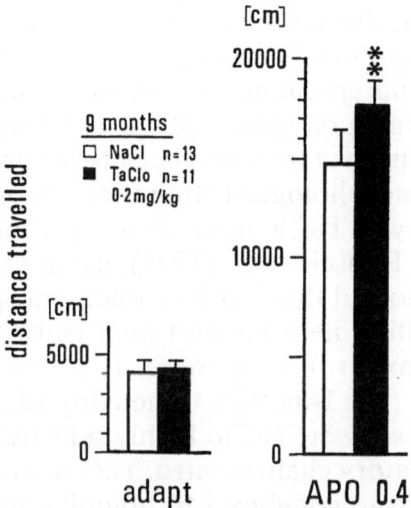

Fig. 3. Distance travelled during the adaptation phase in a new environment 9 months after an injection period of seven weeks with a daily injection of 0.2 mg/kg of TaClo i.p. (left), and after 0.4 mg/kg of apomorphine s.c (right). Statistics: group differences ADAP: $F_{(1,22)} = 9.07$, $p = 0.0028$; ANOVA

Discussion

In summary, these studies with the new neurotoxic-acting substance TaClo show that it induces parkinson-like symptoms, with hyperactivity in the early stages, shortly after the end of the exposure period, and several weeks later a diminished locomotion during the nocturnal active phase and an increased sensitivity to apomorphine in the aging rat 9 months after subchronic exposure to TaClo. Apomorphine appears to interact with up-regulated hypersensitive postsynaptic dopamine receptors.

The neurodegenerative potency of TaClo in rats has recently also been assessed at the higher dose level of 2 mg/kg apomorphine subcutaneously twelve weeks after the daily injection of 0.2 mg/kg TaClo i.p. over a seven week period (Sontag et al., 1995). Because of the induced stereotypic behaviour, rats spent significantly less time running and covered a shorter distance over a 60 minute observation period.

The progressive neurotoxicity of TaClo has also been demonstrated by Grote et al. (1995). These authors used pulse voltammetric measurements to show that intranigral application of TaClo affects the striatal dopamine metabolism of rats one week after application, and that the effect was even more pronounced when analysed three weeks after the injection of TaClo.

It seems that the first signs of dopaminergic denervation are observable 4–9 days after the end of the injection period. In the habituation and exploration phase, rats treated with TaClo were more active, ran faster and covered a longer distance, as described in the present work and also in the recently published data of Sontag and co-workers (1995).

Pulse voltammetric analysis of free-running rats showed an increased release of striatal dopamine during running (O'Neill and Fillenz, 1985). One may therefore assume that during the exploration phase in rats treated with TaClo, released dopamine interacts with up-regulated postsynaptic dopamine receptors. In contrast to this spontaneous behaviour, however, in both these studies the action of 0.4 mg/kg of apomorphine 4–9 days after the application period leads to a shorter distance travelled because of pronounced stereotypic licking behaviour. Histo-morphological studies are in preparation.

The neurotoxic potency of TaClo in in vitro experiments could be demonstrated in several studies. Rausch et al. (1995), using primary cell cultures of mouse mesencephalon, showed that TaClo induces morphological changes in tyrosine hydroxylase-positive neurons and glial cells, e.g. disintegration of dendrites and the loss of axons. Furthermore, Janetzky et al. (1995), demonstrated that as with MPP^+ (the 1-methyl-4-phenylpyridinium ion) but already at a concentration 20 times lower, TaClo highly selectively inhibits complex I of the mitochondrial repiratory chain in vitro. TaClo, as well as MPP^+, inhibits the transfer of electrons from complex I to ubiquinone. These experimental data are comparable with autopsy material from patients with Parkinson's disease. Mizuno et al. (1989), Schapira et al. (1989), Lestienne et al. (1990) and Janetzky et al. (1994) have also described decreased complex I activity in their investigation of the material from the substantia nigra pars compacta.

The behavioural experiments show that the subchronic treatment of rats with a daily low dose of TaClo intraperitoneally for a period of seven weeks leads to significant behavioural changes. Low doses of apomorphine initially stimulate stereotypic movements such as licking, and 9 months after the last injection, a more pronounced effect on locomotor activity as evidenced by the distance travelled. These results lead to the tempting proposition that TaClo induces a slow progressive neurodegenerative process involving the nigrostriatal dopaminergic system.

In the clinical field, attention has been particularly directed towards the neurotoxic potential of trichloroethylene exposure in the work place. Exposure to the industrial solvent trichloroethylene, or solvent abuse with trichloroethylene, or the chronic use of chloral hydrate can lead to endogenous TaClo formation. It cannot be ruled out, therefore, that the damaging effect of TaClo might be one factor responsible for the pathogenesis of environmentally-induced parkinsonism.

Acknowledgement

This study was supported by the Bundesministerium für Bildung, Wissenschaft, Forschung und Technologie, No. 01 KL 9191/0.

References

Bringmann G, God R, Feineis D, Wesemann W, Riederer P, Rausch WD, Reichmann H, Sontag KH (1995) The TaClo concept: 1-trichloromethyl-1,2,3,4-tetrahydro-β-carboline (TaClo) a new toxin for dopaminergic neurons. J Neural Transm [Suppl] 46: 235–244

Bruckner JV, Davis BD, Blancato JN (1989):Metabolism, toxicity and carcinogenicity of trichloroethylene. Crt Rev Toxicol 20: 31–50

Grote C, Clement HW, Wesemann W, Bringmann G, Feineis D, Riederer, P, Sontag KH (1995) Biochemical lesions of the nigrostriatal system by TaClo (1-trichloromethyl-1,2,3,4-tetrahydro-β-carboline) and derivatives. J Neural Transm [Suppl] 46: 275–281

Janetzky B, God R, Bringmann G, Reichmann H (1995) 1-trichloromethyl-1,2,3,4-tetrahydro-β-carboline, a new inhibitor of complex I. J Neural Transm [Suppl] 46: 265–273

Lestienne P, Nelson J, Riederer P, Jellinger K, Reichmann H (1990) Normal mitochondrial genome in brain from patients with Parkinson's disease and complex I defect (published erratum appears in J Neurochem [1991] 56[4]: 1457). J Neurochem 55: 1810–1812

Mizuno Y, Ohta S, Tanaka M, Takamiya S, Suzuki K, Sato T, Oya H, Ozawa T, Kagawa Y (1989) Deficiencies in complex I subunits of the respiratory chain in Parkinson's disease. Biochem Biophys Res Commun 163: 1450–1455

Mohanakumar KP, de Bartolomeis A, Wu RM, Yeh KJ, Sternberger LM, Peng SY, Murphy DL, Chiuh CC (1994) Ferrous-citrate complex and nigral degeneration: Evidence for free radical formation and lipid peroxidation. Ann NY Acad Sci 738: 392–399

O'Neill RD, Fillenz M (1985) Simultaneous monitoring of dopamine release in rat frontal cortex, nucleus accumbens and striatum: effects of drugs, circadian changes and correlations with motor activity. Neuroscience 16: 49–55

Rausch WD, Abdel-mohsen M, Koutsilieri E, Chan WW, Bringmann G (1995) Studies of the potentially endogenous toxin TaClo (1-trichloromethyl-1,2,3,4-tetrahydro-β-carboline) in neural and glia cell cultures. J. Neural Transm [Suppl] 46: 255–263

Schapira AH, Cooper JM, Dexter D, Jenner P, Clark JB, Marsden JD (1989) Mitochondrial complex I deficiency in Parkinson's disease (letter) (see comments). Lancet 1: 1269

Sontag KH, Heim C, Sontag TA, God R, Reichmann H, Wesemann W, Rausch WD, Riederer P, Bringmann G (1995) Long-term behavioural effects of TaClo (1-trichloromethyl-1,2,3,4-tetrahydro-β-carboline) after subchronic treatment in rats. J Neural Transm [Suppl] 46: 283–289

Authors' address: Dr. Christine Heim, Department of Psychiatry, University of Göttingen, von Siebold-Strasse 5, D-37075 Göttingen, Federal Republic of Germany.

Programmed cell death, apoptosis, necrosis and in between

Chair: Y. Mizuno and P. Riederer

Developmental and genetic regulation of programmed neuronal death

M. Weller, J. B. Schulz, U. Wüllner, P. A. Löschmann, T. Klockgether,
and **J. Dichgans**

Department of Neurology, University of Tübingen, Tübingen,
Federal Republic of Germany

Summary.   Apoptotic neuronal death is a key mechanism that regulates the elimination of neuronal precursor cells during the development of the mammalian brain. The principal action of neurotrophins such as nerve growth factor is probably the suppression of the preexistent machinery of programmed cell death that is readily activated in neurons deprived of neurotrophins. Potassium-mediated neuronal depolarization prolongs neuronal survival in vitro and has become a major model of examining neuronal apoptosis. Apoptosis induced by potassium deprivation triggers a lethal cascade of events that includes specific RNA and protein synthesis, induction of interleukin 1-converting enzyme-like protease activity, and generation of free radicals. Neuronal susceptibility to apoptosis is also regulated by the expression of bcl-2 family proteins. Current research focuses on the significance of these findings for the premature death of adult neurons in human neurodegenerative diseases.

Introduction to terminology: Apoptosis and programmed cell death

The maturation of an organism involves the death of numerous cells in various organs at predetermined phases of development. This type of cell death, which plays a major role in shaping the immune and central nervous system during normal development, is a physiological process and is hence termed *developmental* or *programmed cell death*. If neurons fail to establish functionally relevant synaptic connections, their death ensues in the absence of further external molecular stimuli. The concept of programmed cell death opposes certain types of a "meaningful" cell death to other unscheduled types of cell death which are caused, e.g., by mechanical trauma, inflammation or ischemia. The latter are examples of *accidental cell death* and feature disturbances of energy metabolism, cell swelling, osmotic lysis, degeneration of organelles, activation of catabolic enzymes such as proteases and endonucleases and the release of proinflammatory mediators. Accidental cell death results in necrosis in most instances.

In contrast, most types of programmed cell death are characterized by typical morphological changes referred to as apoptosis (Kerr et al., 1972; Cohen, 1993). These include condensation and fragmentation of chromatin and nuclear remnants as well as cytoplasmic compartmentalization, resulting in membrane blebbing and the formation of apoptotic bodies. The latter are efficiently cleared from the tissue by nonprofessional phagocytes (Savill et al., 1993). This process accounts for the lack of inflammatory responses and macrophage activation despite the high rate of cell death during ontogeny. Apoptotic cell death tends to eliminate solitary cells in a given target tissue whereas accidental cell death mostly affects continuous parts of a tissue.

Most types of apoptotic cell death share the biochemical feature of DNA fragmentation. DNA lesions may be restricted to the formation of large DNA fragments of up to 300 kilobase pairs. The classical nucleosomal size, ladder-like pattern of DNA fragmentation that is almost always observed in lymphoid cells undergoing apoptosis may be lacking in apoptosis of nonlymphoid cells. The typical ladder pattern of apoptotic cell death derives from the preferential cleavage of DNA in nucleosomal linker regions which results in the release of 180 base pair fragments and multiples thereof.

Programmed cell death during ontogeny has been considered an *altruistic* suicidal cell death since it serves the purpose of the organism and since there is evidence for an active participation of the dying cell during most instances of programmed cell death (Oppenheim, 1991; Johnson and Deckwerth, 1993; Raff et al., 1993). Thus, inhibition of RNA and protein synthesis prevents numerous types of apoptosis, notably all well-documented forms of neuronal apoptosis, suggesting that cells induced to undergo apoptosis synthesize mRNA encoded by *killer genes*. These mRNA species are translated into *killer proteins* which execute cellular suicide from within. These observations led to the concept of opposing the *suicide* of apoptosis to the *murder* of accidental cell death. However, it is important to note that inhibitors of RNA and protein synthesis are themselves potent inducers of death in most nonneuronal cells in vitro, presumably because such drugs interfere with the synthesis of proteins essential for survival, such as the *bcl*-2 oncogene product in T cells (Weller et al., 1994a). Moreover, such drugs sensitize many cells to at least two potent endogenous mediators of apoptosis, tumor necrosis factor-α and CD95 ligand (Weller et al., 1994b).

Apoptotic cell death during nervous system development

Half of all neuronal precursor cells die during the development of the mammalian central nervous system by a programmed cell death that requires mRNA and protein synthesis and features apoptotic morphology in most instances. In contrast, the role of apoptotic cell death in neuronal loss in systemic degenerations of the adult central nervous system is still controversial (Bredesen, 1995). The regulation of neuronal survival during ontogeny is currently an area of intense research efforts. It is generally assumed that the generation of a multitude of neuronal precursor cells allows for the flexible adaptation of target cells to their proper innervation, that is, all surviving targets receive afferent input, and neurons that fail to reach the appropriate target are eliminated. Direct evidence in support of this concept comes from the observation that augmentation of target structures for neuronal innervation enhances survival of neuronal precursors whereas removal of the target enhances neuronal death (Johnson and Deckwerth, 1993). Investigations into the mechanisms underlying the neurotrophism of nerve growth factor (NGF) led to the identification of an essential regulatory mechanism for neuronal survival, the synthesis of survival factors for approaching neurons by their specific target structures. Factors like NGF are released locally and mediate their effects by acting on specific receptors with tyrosine kinase activity. They maintain neuronal viability and allow for proper synapse and neuronal network formation. In support of the critical role of such factors, neurons deprived of their NGF-releasing target are rescued from apoptosis by exogenous NGF, and the same neurons die in the *presence* of their target when coexposed to neutralizing antibodies to NGF (reviewed in Johnson and Deckwerth, 1993). To consider factors like NGF, brain-derived neurotrophic factor (BDNF) and the neurotrophins (NT) 3 and 4/5 as neurotrophins is misleading in that their principal action is not some ill-defined trophism but to suppress an endogenous cascade of self-destruction and thus to allow synaptic network formation. Incidently, the low affinity NGF receptor, $p75^{NGFR}$, belongs to the same family of cytokine receptors as tumor necrosis factor-receptor and CD95 which are potent mediators of cell death rather than survival. Unexpectedly, forced expression of $p75^{NGFR}$ in the *absence* of NGF promotes neuronal death (Rabizadeh et al., 1993).

The specificity of soluble factors released and their corresponding cell surface receptors seems to allow for a meaningful specific formation of connections between two neuronal systems even over rather long distances and prevents false innervation patterns. Thus, NGF is a specific survival factor for sympathetic neurons, some sensory neurons, and the cholinergic neurons of the basal forebrain that are lost early in the course of Alzheimer's disease. The introduction of the *knock out*-technique (targeted gene disruption) has significantly contributed to our understanding of the role of various neurotrophins in the development of the mammalian brain and has yielded several rather unexpected results which are beyond the scope of this article (for details, see Snider, 1994). However, there are multiple neuronal populations for which no specific survival factor has been identified yet, including neurons with clinical significance such as the dopaminergic neurons of the substantia nigra or the Purkinje cells of the cerebellum, which are lost in Parkinson's disease and some cerebellar ataxias, respectively. The degeneration of the ipsilateral *substantia nigra* after lesioning of the early postnatal striatum (Macaya et al., 1994) provides evidence for a survival factor for dopaminergic neurons, such as glial-derived-neurotrophic factor (GDNF), that is synthesized in the striatum at least during development. Of note, that certain neurotrophins are neuroprotective in experimental models of human neurodegenerative disease, e.g., the MPTP model of Parkinson's disease, is no evidence for a critical role of such factors during ontogenesis. Whether human *adult* neurons also depend on such survival factors, is unknown. Yet, cortico-steroids may be critical survival factors for the granule neurons of the adult hippocampal dentate gyrus since these neurons still undergo apoptosis after adrenalectomy in the adult rat (Sloviter et al., 1993).

Synapse formation and electrical activity promote neuronal survival

The release of survival factors such as the neurotrophins by target structures of neuronal projections is an important but almost certainly not the sole mechanism that determines whether a neuronal precursor cell will die or survive. Thus, certain neuronal populations are susceptible to apoptotic death triggered by removal of afferent input or simply by depressing afferent synaptic activity (Johnson and Deckwerth, 1993; Raff et al., 1993). Intermittent synaptic activity triggers intermittent depolarization and associated calcium fluxes through voltage-dependent calcium channels which appear to be essential for neuronal survival. One of the principal models for the analysis of neuronal survival and apoptosis in vitro that we are studying in our laboratory is based on the survival-promoting effect of chronic potassium-mediated depolarization in cultures of cerebellar granule neurons (D'Mello et al., 1993; Yan et al., 1994, Schulz et al., 1996). In addition to interactions with neuronal and nonneuronal target structures and afferent input, further signals involved in the regulation of developmental neuronal survival may include hormones and astrocyte-derived factors.

Potassium-mediated depolarization blocks apoptosis of cerebellar granule neurons

It has been known since the early 70ies that an elevation of the extracellular potassium concentration greatly prolongs survival of neurons explanted into cell culture dishes. The survival-promoting activity of potassium depends on its depolarizing effect and appears to be mediated by the facilitation of calcium fluxes through voltage-dependent calcium channels. Several types of primary neuronal cultures respond to this effect of potassium-mediated depolarization. In embryonic rat cortical neurons, calcium influx induces the synthesis of BDNF (Ghosh et al., 1994). Furthermore, elevations of intracellular cAMP levels may mimic the effects of chronic potassium depolarization. However, only recently has it been recognized that the survival-promoting effect of potassium-dependent depolarization in cerebellar granule neurons is a powerful model to study programmed neuronal cell death since depolarization enhances survival by suppressing the endogenous pathway of programmed cell death. The latter is inevitably activated after explanting neuronal cells into nondepolarizing culture conditions. Although this system of chronic depolarization is somewhat artificial, it has been considered to mimic a state of differentiation and continuous synaptic activity, reflecting the life of mature neurons in a postnatal normal mammalian brain (Schulz et al., 1996). This assumption is supported by the finding that subtoxic concentrations of the N-methyl-D-aspartat (NMDA) type glutamate receptor agonist NMDA promote cerebellar granule neuron survival *in the absence* of depolarizing potassium concentrations (Yan et al., 1994). Further, a broad screening of cytokines and neurotrophins in the same model system indicated that some cytokines like ciliary neurotrophic factor (CNTF), tumor necrosis factor-α and the interleukins 10 and 13 promote survival of nondepolarized cerebellar granule neurons whereas other cytokines such as transforming growth factor-β and BDNF precipitated premature neuronal death (De Luca et al., 1996a, 1996b).

Apoptotic developmental neuronal death is an active type of cell death

The antiapoptotic effects of inhibitors of RNA and protein synthesis such as actinomycin D and cycloheximide in probably all instances of neuronal apoptosis examined so far have provided strong evidence for an active participation of neurons during their programmed death (Johnson and Deckwerth, 1993). This has provoked an intense search for genes and proteins that are involved in the execution of neuronal cell death, *killer genes* and *killer proteins* (Schwartz and Osborne, 1993), the most important of which are shown in Table 1. For most of these genes and proteins, however, it has not been demonstrated that their induction during apoptosis is critical to the death process. Indeed, enhanced expression of some of these factors like c-Fos or Hsp-70 might represent an effort by the neuron to counteract apoptosis. The killer proteins might be nucleases or enzymes themselves and participate directly in the execution of death or may be transcription factors controlling the expression of such enzymes on the genetic level or may act by activating or stabilizing preexisting death proteins (Johnson and Deckwerth, 1993). We are currently exploring the role of some of these candidate proteins in neuronal apoptosis triggered by potassium deprivation. Preliminary evidence suggests that c-Jun but not c-Fos is induced in cerebellar granule neurons forced into apoptosis (Schulz et al., unpublished observation). The role of factors with presumptive significance for neuronal survival in vitro can subsequently be

Table 1. Putative inducers and inhibitors of neuronal apoptosis

Candidate killer genes and proteins	*Antiapoptotic genes and proteins*
c-jun	bcl-2
c-fos	bcl-x
sulfated glycoprotein-2	growth factor receptors
tissue transglutaminase	
RP2	
RP8	
ICE-like proteases	
bax	
p53	

examined in animal models of human neurodegenerative disorders, e.g., in the naturally occurring mouse mutants *weaver* and *lurcher*. These mice exhibit a developmentally regulated pattern of enhanced apoptosis of cerebellar granule neurons and Purkinje cells, respectively (Wüllner et al., 1995).

Prevention of apoptosis by inhibitors of protein synthesis may not always prove that inhibition of killer protein synthesis is involved. An alternative pathway for cycloheximide-mediated prevention of neuronal apoptosis may involve the shift of cystein from protein synthesis to glutathione synthesis with a consecutive increase in antioxidative properties (Ratan et al., 1994).

That specific genes control the rate of programmed cell death not only in the central nervous system, has been demonstrated in the nematode, *C. elegans*. This worm loses 131 of its 1090 cells during development in a process that is tightly regulated by two proapoptotic genes, *ced*-3 and *ced*-4, and one antiapoptotic gene, *ced*-9. *Ced*-9 is the homolog of the human *bcl*-2 gene, as confirmed by gene transfer studies which have shown that a *bcl*-2 gene transfer can compensate for a targeted loss of the *ced*-9 gene in *C. elegans*. A mammalian homolog for *ced*-4 awaits to be identified. *Ced*-3 is homologous to the mammalian family of interleukin 1-converting enzyme-(ICE)-like proteases (Kumar and Harvey, 1995). Activation of such enzymes appears to be involved in potassium deprivation-induced apoptosis of cerebellar granule neurons, too (Schulz et al., 1996). Relatively little is known about the events downstream of ICE activation or activation of other proteases or nucleases and specifically about the proximate course of neuronal death. Recent studies

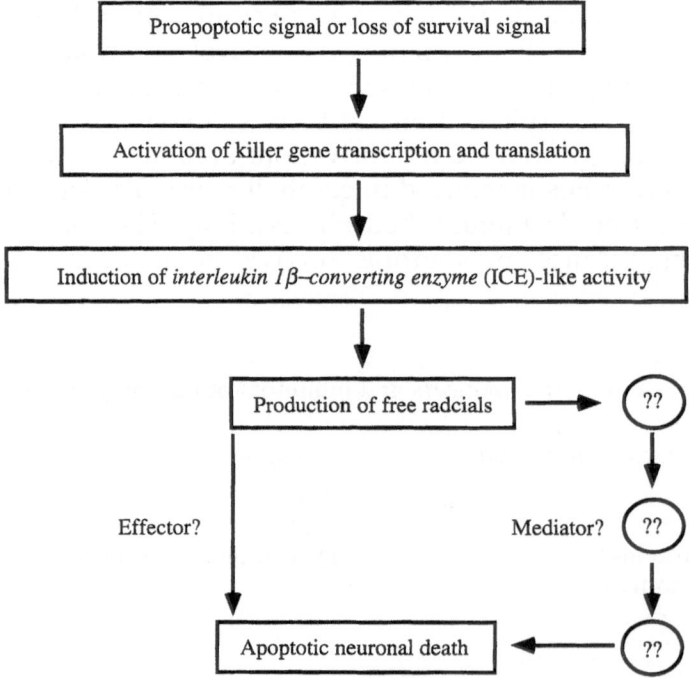

Fig. 1. A putative cascade of potassium deprivation-induced apoptosis of cerebellar granule neurons (for details, see Schulz et al., 1996)

from our laboratory and other groups have defined an important role for reactive oxygen species as down-stream mediators or effectors of neuronal apoptosis but how these reactive intermediates eventually kill a cell remains still unclear (Bredesen, 1995; Schulz et al., 1996).

The *bcl*-2 family of apoptosis-regulating genes

The antiapoptotic protooncogene *bcl*-2 was detected in follicular lymphomas because of a translocation of the *bcl*-2 gene from chromosome 18q21 to chromosome 14q32 where it came under the transcriptional control of strong enhancer regions for the immunoglobulin heavy chain gene. Subsequent analyses showed that multiple other neoplasms express enhanced levels of Bcl-2 protein in the absence of such translocations. Bcl-2 inhibits multiple forms of apoptotic cell death, including apoptosis after growth factor deprivation, serum starvation, irradiation and exposure to cancer chemotherapy drugs and cytotoxic cytokines (Reed, 1994). It has been suggested that *bcl*-2 enhances the cellular antioxidative properties. However, this cannot account for all instances of *bcl*-2-mediated rescue from apoptosis since *bcl*-2 blocks apoptosis in oxygen-free conditions. *Bcl*-2 is a prototype member of a family of genes involved in the control of cell death. Some of the gene products such as Bcl-x inhibit apoptosis, others such as Bax promote cell death. *Bcl*-2 family genes are highly expressed during the development of the central nervous system and some neurons retain expression of Bcl-2 or Bcl-x in adulthood. While the expression of Bcl-2 in non-neuronal cells such as T lymphocytes or myeloid cells is tightly controlled by survival factors such as interleukin-2 or-3 (Weller et al., 1994a, 1995), the molecular mechanisms that regulate expression of *bcl*-2 and related genes in neurons are largely unknown. Nevertheless, forced expression of *bcl*-2 blocks neuronal death precipitated by survival factor deprivation in factor-dependent neurons, e.g., sympathetic neurons deprived of NGF or sensory neurons deprived of BDNF or NT-3, as well as death of neuronal cell lines triggered by glucose deprivation, serum deprivation, calcium ionophore or free radicals (Garcia et al., 1992; Allsopp et al., 1993; Kane et al., 1993). There might be, however, a certain specificity to the antiapoptotic effects of *bcl*-2 in neuronal cells since CNTF deprivation-induced death of dependent ciliary neurons is not blocked by *bcl*-2 (Allsopp et al., 1993). Further, there was an unexpected reduction in *bcl*-2 mRNA expression in cerebellar granule neurons exposed to survival-promoting concentrations of NMDA (Montpied et al., 1993), and the antiapoptotic properties of cytokines such as CNTF, tumor necrosis factor-α and the interleukins 10 and 13 in the same neurons do not involve induction of *bcl*-2 expression (De Luca et al., 1996b).

Concluding remarks

Half of all neuronal precursor cells die during the normal development of the central nervous system in a process of programmed cell death that is genetically controlled and exhibits features of an active type of cell death. The molecular mechanisms underlying programmed neuronal death are currently being characterized, and significant progress in our understanding of these processes can be expected within the next years. While these issues are of general interest to those interested in developmental neurobiology, there is also significant hope that understanding programmed developmental neuronal death will also result in a better understanding of and eventually new treatment strategies for various neurodegenerative disorders including Parkinson's disease, Alzheimer's disease, Huntington's disease, hereditary and idiopathic cerebellar ataxias and motor neuron disease, as well as infectious diseases such as Prion- and HIV-associated neurological disorders and, finally, ischemic brain damage.

References

Allsopp TE, Wyatt S, Paterson HF (1993) The proto-oncogene *bcl*-2 can selectively rescue neurotrophic factor-dependent neurons from apoptosis. Cell 73: 295–307

Bredesen DE (1995) Neural apoptosis. Ann Neurol 38: 839–851

Cohen JJ (1993) Apoptosis. Immunol Today 14: 126–131

De Luca A, Weller M, Fontana A (1996a) TGF-β-induced apoptosis of cerebellar granule neurons is prevented by depolarization. J Neurosci (in press)

De Luca A, Weller M, Frei K, Fontana A (1996b) Maturation-dependent modulation of apoptosis in cultured cerebellar granule neurons by cytokines and neurotrophins. Eur J Neurosci 13: 4174–4185

D'Mello SR, Galli C, Ciotti T, Calissano P (1993) Induction of apoptosis in cerebellar granule neurons by low potassium: inhibition by insulin-like growth factor I and cAMP. Proc Natl Acad Sci USA 90: 10989–10993

Garcia I, Martinou I, Tsujimoto Y, Martinou JC (1992) Prevention of programmed cell death of sympathetic neurons by the *bcl*-2 proto-oncogene. Science 258: 302–304

Ghosh A, Carnahan J, Greenberg ME (1994) Requirement for BDNF in activity-dependent survival of cortical neurons. Science 263: 1618–1623

Johnson EM, Deckwerth TL (1993) Molecular mechanisms of developmental neuronal death. Ann Rev Neurosci 16: 31–46

Kane DJ, Sarafian TA, Anton R, Hahn H, Gralla EB, Valentine JS, Örd T, Bredesen DE (1993) Bcl-2 inhibition of neural death: decreased generation of reactive oxygen species. Science 262: 1274–1277

Kerr JFR, Wyllie AH, Currie AR (1972) Apoptosis: a basic biological phenomenon with wide-ranging implications in tissue kinetics. Br J Cancer 26: 239–257

Kumar S, Harvey NL (1995) Role of multiple cellular proteases in the execution of programmed cell death. FEBS Lett 375: 169–173

Macaya A, Munell F, Gubits RM, Burke RE (1994) Apoptosis in substantia nigra following developmental striatal excitotoxic injury. Proc Natl Acad Sci USA 91: 8117–8121

Montpied P, Weller M, Paul SM (1993) N-methyl-D-aspartate receptor agonists decrease protooncogene *bcl*-2 mRNA expression in cultured rat cerebellar granule neurons. Biochem Biophys Res Commun 195: 623–629

Oppenheim RW (1991) Cell death during development of the nervous system. Ann Rev Neurosci 14: 453–501

Rabizadeh S, Oh J, Zhong LT, Yang J, Bitler CM, Butcher LL, Bredesen DE (1993) Induction of apoptosis by the low-affinity NGF receptor. Science 261: 345–348

Raff MC, Barres BA, Burne JF, Coles HS, Ishizaki Y, Jacobson MD (1993) Programmed cell death and the control of cell survival. Science 262: 695–700

Ratan RR, Murphy TH, Baraban JM (1994) Macromolecular synthesis inhibitors prevent oxidative stress-induced apoptosis in embryonic cortical neurons by shunting cysteine from protein synthesis to glutathione. J Neurosci 14: 4385–4392

Reed JC (1994) Bcl-2 and the regulation of programmed cell death. J Cell Biol 124: 1–6

Savill J, Fadok V, Henson P, Haslett C (1993) Phagocyte recognition of cells undergoing apoptosis. Immunol Today 14: 131–136

Schulz JB, Weller M, Klockgether T (1996) A sequential requirement for new mRNA and protein synthesis, ICE-like protease activity, and free radicals in potassium deprivation-induced apoptosis of cerebellar granule cells. J Neurosci 16: 4696–4706

Schwartz LM, Osborne BA (1993) Programmed cell death, apoptosis and killer genes. Immunol Today 14: 582–590

Sloviter RS, Sollas AL, Dean E, Neubort S (1993) Adrenalectomy-induced granule cell degeneration in the rat hippocampal dentate gyrus: characterization of an in vivo model of controlled neuronal death. J Comp Neurol 330: 324–336

Snider WD (1994) Functions of the neurotrophins during nervous system development: what the knockouts are teaching us. Cell 77: 627–638

Weller M, Constam DB, Malipiero UV, Fontana A (1994a) Transforming growth factor β_2 induces apoptosis of murine T cell clones without down-regulating bcl-2 mRNA expression. Eur J Immunol 24: 1293–1300

Weller M, Frei K, Groscurth P, Krammer PH, Yonekawa Y, Fontana A (1994b) Anti-Fas/APO-1 antibody-mediated apoptosis of cultured human glioma cells. Induction and modulation of sensitivity by cytokines. J Clin Invest 94: 954–964

Weller M, Malipiero UV, Aguzzi A, Reed JC, Fontana A (1995) Protooncogene bcl-2 gene transfer abrogates Fas/APO-1 antibody-mediated apoptosis of human malignant glioma cells and confers resistance to chemotherapeutic drugs and therapeutic irradiation. J Clin Invest 95: 2633–2643

Wüllner U, Löschmann PA, Weller M, Klockgether T (1995) Apoptotic cell death in the cerebellum of mutant weaver and lurcher mice. Neurosci Lett 200: 109–112

Yan GM, Ni B, Weller M, Wood KA, Paul SM (1994) Depolarization or glutamate receptor activation blocks apoptotic cell death of cultured cerebellar granule neurons. Brain Res 656: 43–51

Authors' address: Dr. M. Weller, Department of Neurology, University of Tübingen, Medical School, Hoppe-Seyler-Strasse 3, D-72076 Tübingen, Federal Republic of Germany.

Apoptosis in neurodegenerative disorders

H. Mochizuki, H. Mori, and **Y. Mizuno**

Department of Neurology, Juntendo University School of Medicine, Tokyo, Japan

Summary. Although the exact mechanism of nigral cell death in Parkinson's disease (PD) is not known, increasing evidence suggests the presence of apoptotic cell death in PD. When we applied the TUNEL method to detect DNA fragmentation, four out of seven late onset sporadic patients with PD showed TUNEL-positive neurons. The percentages of those neurons among the remaining melanin containing neurons were 0.6 to 4.8% (average 2.1%). But TUNEL-positive neurons could not be detected in control subjects as well as four patients with young onset (under 40 years of the age) PD. Numbers of nigral toxins such as MPTP, complex I inhibitors, and mitochondrial respiratory inhibitors have been reported to induced apoptotic cell death. These findings suggest that apoptosis is involved in nigral cell cleath in PD at least in part and warrant further studies on apoptosis-related substances in PD.

Introduction

Neurodegenerative disorders are characterized by slowly progressive loss of neurons; each disease has a specific distribution of neural degeneration. Neuronal loss does not take place at one time; one neuron dies after the other very slowly over a long period of time. Neuronal degeneration starts far beyond the clinical manifestation of symptoms; for instance, in case of Parkinson's disease (PD), more than 70% of nigral neurons are said to be lost before the appearance of the initial symptom. The character of neuronal death in neurodegenerative disorders has never been critically discussed until recently. Probably, it has been thought that neurons die by necrosis. Necrosis refers to the morphology most often seen when cells die from severe and sudden injury, such as anoxia, hyperthermia, physical trauma or chemical damage (Cohen, 1993); the plasma membrane loses its ability to regulate osmotic pressure, and the cell swells and its contents are spilled into the surrounding tissue (Cohen, 1993). But the characteristics of neuronal death in neurodegenerative disorders is more protracted. Swollen neurons may be seen in some disorders such as amyotrophic lateral sclerosis and corticobasal degeneration, but in most of the neurodegenerative diseases, neurons shrink. Therefore, it seems more likely that neurons die by apoptosis.

The term apoptosis was introduced by Kerr et al. (1972) based on the morphologic characteristics of dying cells that differed from necrosis. According to Kerr et al. (1972), the first structural changes take place in two discrete stages. The first stage comprises nuclear and cytoplasmic condensation and breaking up of the cell into a number of membrane-bound, ultrastructurally well-preserved fragments. In the second stage these apoptotic bodies are shed from epithelial-lined surfaces or are taken up by other cells. The structural integrity of the plasma membrane and cellular organelles remain intact (Bredesen, 1995). The term apoptosis is often equated with programmed cell death, but they are not exactly the same (Cohen, 1993); as not all the apoptotic cell deaths are programmed.

Programmed cell death was found as predictable death of certain groups of cells at the finite stage of the development (Lockshin and Beaulaton, 1974). In 1964, Lockshin and Williams found death of specific cells in an apparently predetermined sequence at certain stages of insect metamorphosis, and used the term "programmed cell death". The programmed cell death has extensively been studied in the nematode *Caenorhabditis elegans* (Sulston and Horvitz, 1977) and it has been found that three genes, two proapoptotic genes *ced*-3 and *ced*-4 and one antiapoptotic gene *ced*-9, regulate the programmed cell death of *C. elegans* (Yuan and Hortita, 1990; Yuan et al., 1993; Hengartner and Horvitz, 1994). Morphologic characteristics of programmed cell death mostly conform to those of apoptosis, although not all the programmed cell deaths show morphologic characteristics of apoptosis (Clarke, 1990). It has been shown that apoptosis is seen not only in development but also in normal turnover of cells, ageing and pathology (Kerr et al., 1972; Cohen, 1993; Bredesen, 1995). At the biochemical level, apoptosis is associated with protein synthesis and DNA fragmentation in the internucleosomal regions (Bredesen, 1995; Wyllie, 1980). The subject of apoptosis has extensively been reviewed recently by Bredesen (1995). We have been interested in whether or not apoptosis is involved in neurodegenerative disorders, and we looked at MPP+-induced cell death in culture and then the substantia nigra of Parkinson's disease for apoptosis. Details of our studies have been reported elsewhere (Mochizuki et al., 1994, 1996).

MPP+-induced apoptosis

We established a co-culture of the ventral mesencephalon and the neostriatum from rat embryos at embryonic day 14. The neostriatum and ventral mesencephalon were dissected in cold calcium- and magnesium-free Hank's balanced salt solution and were grown in minimum essential medium (MEM)/ F12 with 10% fetal bovine serum. Thereafter, cells were treated with MEM/ F12 and 5% newborn calf serum/5% horse serum without using cytosine arabinoside.

Cultured tissues were incubated with a medium containing 20 μM MPP+ or a vehicle for 48 hours starting Day 5 of culture. Morphological changes were analyzed by immunocytochemistry for tyrosine hydroxylase (TH) and by the

TUNEL (terminal deoxynucleotidyltransferase nick-end labeling) method. TH immunohistochemistry was performed according to the standard ABC method as reported previously (Mochizuki et al., 1994). The tissue specimens were lightly stained with hematoxylin to visualized nuclear structures after immunostaining for TH.

For visualizing apoptosis, we used the TUNEL method. Mesencephalic culture was treated with biotinated deoxyuridine triphosphate in the presence of terminal deoxynucleotidyl transferase (TdT) by incubating in TDT buffer (30 mM tris-HCl, 140 mM sodium cacodylate, and 1 mM cobalt chloride) at 37°C for 1 hr. Biotinated deoxyuridine attached to the 3′-end of nucleotides of nucleosomes was visualized by incubation with streptavidin peroxidase complex for 1 hr and then by diaminobenzidine.

For DNA electrophoresis, cultured cells were lysed by adding 0.5 ml of 10 mM Tris-HCl buffer pH 8.0 and 1 mM EDTA and cells were detached with a cell scraper. DNA was extracted with ethanol/chloroform/isoamyl alcohol and recovered by centrifugation. Equal amounts of DNA were then loaded on 1.8% agarose gels containing ethidium bromide and photographs were taken on a UV transilluminator.

In the TH immunostaining of normal culture, only a small number of neurons showed nuclear chromatin condensation. In the culture treated with 20 µM of MPP⁺, marked nuclear chromatin condensation around TH-positive neurons was seen.

In TUNEL of culture treated with MPP⁺, many nuclei were positively stained. Not only the neuronal nuclei but also glial nuclei were positively stained. In addition, probably non-TH-positive neurons were also positively stained. In the control culture, only a small number of nuclei were positively stained.

In the DNA electrophoresis of the culture treated with MPP⁺, typical DNA ladder was observed.

In the literature, Dipasquale et al. (1991) reported apoptosis of cultured cerebellar granular neurons by MPP⁺, and Hartley et al. (1994) and Itano et al. (1994) reported induction of apoptosis in PC12 cells by MPP⁺.

Apoptosis in Parkinson's disease

Then we studied the substantia nigra of patients died of PD. The clinical subjects studied include 6 control patients, 4 young-onset PD patients, and 7 late onset PD patients. The age of the patients and the postmortem time are shown in Table 1. The young onset PD patients had the onset of the disease before 40 years of the age. All of the PD patients studied had typical pathologic changes of PD consisting of degeneration of the substantia nigra and the locus coeruleus with Lewy bodies in the remaining neurons.

Six µM-thick paraffin-embedded transverse midbrain sections were deparaffinized with xylene and used for nick-end labeling. Endogenous peroxidases were inactivated with 2% H₂O₂ for 5 minutes. The sections were rinsed, and immersed in TDT buffer. Sections were incubated with terminal deoxynucleotidyl transferase (TdT) and biotinated deoxyuridine triphosphate (dUTP) in

Table 1. TUNEL in Parkinson's disease

No	Dx	Age	DD	PMT	% (+)
1–6	C	18 ~ 99		1 ~ 3	0
7–10	Y	50 ~ 72	14 ~ 33	3 ~ 12	0
11	L	76	18	3	4.2
12	L	78	13	2	2.1
13	L	61	16	5	0
14	L	75	3	3	1.2
15	L	81	8	2	0.6
16	L	66	10	3	0
17	L	75	10	1	0

C control, *Y* young onset PD, *L* late onset PD, *DD* disease duration, *PMT* postmortem time in hours, % (+) % of TUNEL-positive neurons in the SN. Cited from Mochizuki et al., 1996

TDT buffer for 60 minutes at 37°C. Sections were rinsed and covered with streptavidine peroxidase complex for one hour at room temperature. Then labeling was visualized by incubating with diaminobenzidine.

Both positive controls and negative controls were run together. In the positive control, specimens were pretreated with DNAase I. Therefore, in this specimen, most of the nuclei should be labeled positively. In the negative control, TdT was omitted from the incubation mixture.

Practically, all the nuclei were positively labeled when sections were pretreated with DNAase I, and no staining was observed when TdT was omitted. In all the control subjects studied, no nigral pigmented neurons were positively stained by TUNEL. No TUNEL-positive neurons were found in other areas of the midbrain either. Only one of the control subjects had TUNEL-positive glial nuclei in the substantia nigra.

In young onset PD patients, no nigral or other neurons were positively stained by TUNEL. In late onset PD patients, three out of seven patients showed TUNEL-positive neurons in the substantia nigra. These were melanin containing neurons. However, the numbers of TUNEL-positive neurons were only few, from 0.6 to 4.8%, average 2.1% (Table 1). Some TUNEL-positive glial nuclei were found in 6 out of the 7 late onset PD patients and in 2 out of the 4 young onset patients.

No correlation was noted between the TUNEL-positivity and the postmortem time; in fact, one of the young-onset PD patients had the longest postmortem time; still no TUNEL-positive neurons were seen. As the number of TUNEL-positive cells was small, it was difficult to analyze the correlation between the disease duration or severity with TUNEL-positivity. No apparent correlation was noted between the age of the patients and TUNEL-positivity. Actually, the eldest person in the control group (99 years of the age) did not show TUNEL-positive neurons.

In young-onset PD patients, no nuclei were positively stained. This may be due to much slower degeneration of nigral cells compared to late onset PD. It has been claimed that the neurons showing nuclear fragmentation are going to die within 5 hours (Cohen et al., 1991). Therefore, chances are small for nuclear fragmentation to be picked up in autopsy materials by this method. Fragmented DNAs are quickly taken up by phagocytes and disappear by digestion. Therefore, even a small proportion, positive labeling would indicate the involvement of apoptotic process in the substantia nigra of PD. Specificity of the TUNEL method will be discussed later.

In the literature, Dragunow et al. (1995) studied three PD patients but no TUNEL-positivity was observed. Recently, two other groups reported apoptosis in PD. Anglade et al. (1995) studied 3 parkinsonian brains and 3 control brains at the ultrastructural level; they found chromatin condensation and cell shrinkage which would indicate apoptosis in 6 out of 160 nigral neurons, but they did not find such neurons among 90 neurons examined on control brains. Tompkins and Hill (1995) studied 28 brains including PD, AD/PD, DLBD as well as control diseases; they found TUNEL-positive reaction in the nigral neurons not only in PD and DLBD brains but also in control and AD brains; the percentages of positive staining was reported to be higher in PD. They suggested that in PD, AD/PD, and DLBD, neurons would die by apoptosis. Therefore, apoptosis appears to be involved at least in part in nigral neuronal death in PD.

Apoptosis in other neurodegenerative disorders

Then, the next question is whether or not apoptosis is involved in neural cell death in other neurodegenerative disorders. Neural death takes place very slowly in these diseases. Therefore, apoptosis may be the final common pathway for cell death despite the difference in pathogenesis. There are several reports already in the literature on other neurodegenerative disorders.

Su et al. (1994) reported numerous neuronal nuclei displaying distinct morphological characteristics of apoptosis within tangle-bearing neurons as well as non-tangle-bearing neurons in AD brains by the TUNEL method, whereas they saw few or no such nuclei in control brains. Lassmann et al. (1995) also analyzed 18 AD and 15 age-matched control brains by the TUNEL method. They observed 30 times more brain cells showing DNA fragmentation in AD as compared to age-matched controls; approximately 25% of neurons were positively stained, and neurons situated within areas of amyloid deposits or affected by neurofibrillary degeneration revealed a higher risk of DNA fragmentation. According to their data, not only neurons but also oligodendrocytes and microglia were positively stained. In AD, β-amyloid peptide may trigger apoptosis, because cultured neurons treated with β-amyloid peptides exhibited morphological and biochemical characteristics of apoptosis (Forloni et al., 1993; Loo et al., 1993).

In Huntington's disease, Portera-Cailliau et al. (1995) studied 13 autopsied brains by the TUNEL method; they reported that striatal neurons were positively stained in 2 out of the 3 patients in grade 2, 4 out of the 5 patients in grade 3, and 1 out of 4 patients in grade 4; they also observed labeling of oligodendrocytes but not neurons in 1 patient in grade 0. They ascribed oligodendroglial labeling to either Wallelian degeneration of axons from neuronal death resulting in secondary cell death of oligodendrocytes or to a direct result of the genetic alteration in the IT-15 gene on oligodendroglia. Thus not all the patients whom they studied showed TUNEL-positivity, and there was no direct correlation between the grade of the disease and positive labeling; furthermore, there was no correlation between the postmortem time and the TUNEL-positivity. Dragnow et al. (1995) also studied Huntington's patients and 4 out of the 7 patients showed TUNEL-positivity in striatal neurons.

What may be the mechanism of TUNEL-positivity in Huntington's disease? Quinolic acid-induced experimental biochemical model of Huntington's disease in rats also showed extensive labeling of striatal neurons (Portera-Cailliau et al., 1995); the dorsal putamen showed much higher densities of TUNEL-positivity; they also observed DNA laddering in the earlier stage but DNA smear in the later stage in the quinolinic model, and postulated that both apoptotic and necrotic processes might be taking place simultaneously.

In ALS, Yoshiyama et al. studied 10 ALS patients using TUNEL and Le^Y antigen immunohistochemistry; Le^Y is a difucosylated type 2 chain determinant which has been identified as a characteristic of cells undergoing apoptosis (Hiraishi et al., 1993). In their study, 4 out of 10 ALS patients showed TUNEL-positive neurons in the anterior horns, however, the number of positively-stained neurons were only one or two among 15 to 62 anterior horn cells counted. In the Le^Y immunohistochemistry, 7 out of 10 patients showed Le^Y-positive neurons; the numbers of positively-stained neurons were 1 to 5 among 5 to 62 neurons counted.

More direct evidence to indicate the involvement of apoptosis in neurodegenerative disorders came from recent observation on spinal muscuiar atrophy. Spinal muscular atrophies (SMAs) are a group of autosomal recessive disorder affecting the anterior horn cells; type I (Werdnig-Hoffmann's disease), type II, and type III (Kugelberg-Welander disease) are subclassified based on the age of onset and severity of muscle atrophy (Munsat, 1991). Recently, Roy et al. (1995) found a novel gene for neuronal apoptosis inhibitory protein (NAIP) and they mapped the SMA region to chromosome 5q13.1; the disease phenotypes of SMAs were associated with deletion mutations of this gene. Furthermore, they showed significant homology of NAIP with the two baculoviral gene products, AplAP and OplAP, which are capable of inhibiting insect cell apoptosis (Clem and Miller, 1994). NAIP is able to protect apoptosis of cultured cells induced by serum deprivation, free radicals, or by tumor necrosis factor-α (Liston et al., 1996).

Thus nuclear fragmentation appears to be a common phenomenon in various kinds of neurodegenerative disorders.

Postmortem time, age of the patients, and TUNEL-positivity

Lassmann et al. (1995) studied the relationship between the number of cells showing nuclear fragmentation per square mm with the postmortem time and the age of the patients at death by the TUNEL method in Alzheimer's disease and the age matched control subjects; they found neither correlation between the number of positive cells and the age of the patients at death nor between the number of positive cells and the postmortem time.

In HD, Portera-Cailliau et al. (1995) also found no correlation between the relative number of TUNEL-positive cells and the age, postmortem time or duration of the disease; their eldest patient who was 84-year-old showed 1+ positivity and a 76-year-old patient showed 3+ positivity; the longest postmortem time 14 hours was seen in 77-year-old patient who showed no TUNEL-positive cells; on the other hand, the above 76-year-old patient with 3 hours postmortem time showed 3+ positivity.

In PD in our hands, again no correlation was noted between the post mortem time and TUNEL-positivity; a 78-year-old patient with 2 hours post mortem time showed TUNEL-positive cells; on the other hand, a 61-year-old patient with 5 hours postmortem time did not show positive neurons. Also, no correlation was noted between the TUNEL-positivity and the age of the patients at death or duration of the disease. Particularly, a 67-year-old young onset PD patient with 30 years history and 12 hours postmortem time did not show TUNEL-positive cells indicating that neither the duration of the disease nor postmortem delay would significantly influence the TUNEL-positivity.

Other reports also support lack of correlation between the TUNEL-positivity and postmortem time (Su et al., 1994; Dragnow et al., 1995; Yoshi-yama et al., 1995). None of the reports addressed the possible influence of ante-mortem conditions on TUNEL-positivity. The antemortem conditions vary so much among patients and it is by no means easy to evaluate this question quantitatively. However, in the above reports, it is quite rare for control subjects to show TUNEL-positivity. Control subjects would also suffer from antemortem hypoxic and hypotensive episodes. Therefore, again influence from antemortem conditions appears to be small.

Specificity of the TUNEL method

One of the important questions to be answered on TUNEL method is whether or not TUNEL differentiate apoptosis from necrosis. Recently Grasl-Kraupp et al. (1995) reported that the TUNEL assay failed to discriminate among apoptosis, necrosis and autolytic cell death; they observed TUNEL-positive reaction in necrotic hepatocytes after a cytotoxic dose of carbon tetrachloride or N-nitrosomorpholine and in insufficiently fixed autolytic livers. They conclude that DNA fragmentation is common to different kinds of cell death. But their situation is a very artificial one giving massive dose of hepatotoxic substances to rats. Situation in the central nervous system of humans may not be the same as liver cells of rats treated with hepatotoxins. First of all, in postmortem human brains, autolytic process is an inevitable process before fixation of brain in formalin. Furthermore, antemortem agonal process may be more disastrous to neurons because situations similar to hypoperfusion-reperfusion or deoxygenation-reoxygenation are very common among sick patients undergoing the morbibound process. Despite these facts, we could not detect TUNEL-positive reaction in control brains as well as in brains of young-onset PD. Furthermore, only small proportions of nigral neurons showed TUNEL-positive reaction in adult-onset PD. In addition, structures other than the substantia nigra in the midbrain were not positively stained. Therefore, in our materials it seems unlikely that TUNEL-positivity indicates necrosis or autolysis.

Recently, Gold et al. (1994) compared TUNEL (use of terminal transferase) with nick translation (use of DNA polymerase); they used in vitro (apoptosis of thymocytes induced by γ-irradiation and necrosis of thymocytes induced by the cytotoxic action of antibody and complement) and in vivo (apoptosis of autoimmune encephalomyelitis and necrosis of kainic acid-induced nerve cell degeneration) models for apoptotic or necrotic dell death. In their study, DNA fragmentation was visualized by the incorporation of labeled nucleotides into the nuclei of affected cells utilizing tailing (use of terminal transferase) or nick translation techniques. They observed preferential labeling of cells undergoing apoptosis by tailing in the early stages of cell degeneration in vitro, whereas necrotic cells were identified by nick translation. They report similar results in in vivo models.

Thus it appears to be that TUNEL-positivity indicates probable presence of apoptotic cell death, but one could not be quite sure whether or not some of the TUNEL-positive cells are undergoing necrosis. Probably, simultaneous examination by TUNEL and nick translation would give more definite results.

But in neurodegenerative disorders, strict differentiation of neural death whether it is apoptosis or necrosis may not be so important. Whichever the process is, neural death progresses very slowly, and apoptosis and/or necrosis appear to be the final process of cell death. Before reaching that end point, numbers of biochemical abnormalities must have taken place in dying neurons including upregulation or down regulation of genes regulating cell survival and cell death. Disclosure of those processes are more important to explore the exact mechanisms of neural death and to develop new methods to rescue those dying neurons. In this regard, studies on genes regulating apoptosis appears to be a very important subject of research in neurodegenerative disorders.

Mechanism of apoptosis in neurodegenerative disorders

Suppose apoptosis is involved in neurodegenerative disorders, what may be the mechanism. In Alzheimer's disease, accumulation of β-amyloid peptide may trigger the apoptotic process. Loo et al. (1993) reported induction of apoptosis by treatment of cortical culture with β-amyloid peptide; in their studies electron microscopic observation of the cultured neurons showed membrane brebbing and compaction of nuclear chromatin; furthermore, agarose gel electrophoresis revealed internucleosomal DNA fragmentation. In this respect, Lassmann et al. (1995) found a weak correlation between the neurofibrillary tangle and the TUNELpositive neurons in Alzheimer's disease, but not with senile plaques where β-amyloid peptide is located. In addition, it has been reported that β-amyloid peptide induces necrosis rather than apoptosis (Behl et al., 1994).

In Huntington's disease, excitotoxicity may be involved in inducing apoptosis. Intrastriatal injection of a glutamate receptor agonist (excitotoxins) recapitulate some neuropathological features of HD (Portera-Cailliau et al., 1995). Potera-Cailliau et al. (1995) observed internucleosomal DNA fragmentation at early time after quinolinate injection, but in the later time points, random DNA fragmentation was observed; they postulated that both apoptotic and necrotic processes might take place simultaneously.

The mechanism of apoptosis in ALS is unclear. Approximately 10 to 15% of ALS patients show familial occurrence with autosomal dominant inheritance (Brown, 1995), and about 20% of familial ALS show mutations in *SOD 1*, (the gene encoding copper/zinc superoxide dismutase, SOD 1) are associated with the disease phenotype (Rosen et al., 1993). The activity of SOD 1 in these patients is not always low; glycine to alanine mutation at codon 93 of exon 4 produces little decrease in the red blood cell SOD 1 activity (Esteban et al., 1994). In addition, transgenic mice expressing this mutation did show motor neuron death (Gurney et al., 1994). Therefore, decrease in the activity per se does not appear to be responsible for the motor neuron degeneration. Recently, Rabizadeh et al. (1995) reported an interesting observation in that the overexpression of wild-type *SOD 1* inhibited apoptosis of cultured neural cells induced by serum and growth factor withdrawal or calcium ionophore; in contrast, familial ALS-associated *SOD 1* mutants promoted neural apoptosis.

In spinal muscular atrophy associated with mutations of NAIP gene, apoptosis may be a direct effect of this apoptosis inhibitory gene and protein as mentioned above (Roy et al., 1995).

Mechanism of cell death in Parkinson's disease

If apoptosis is involved in PD, what may be the mechanism. Numbers of substances such as cyokines, glucocorticoid, Fas ligands, chemotherapeutic agents, and trophic factor deprivation can induced apoptosis. Recently there is a growing evidence indicating the involvement of these factors in neurodegeneration of PD. Basic fibroblast growth factor (bFGF) is a peptide which has a neurotrophic effect on mesencephalic dopaminergic neurons (Engele et al. 1991). It reduces the in vivo neurotoxic effects of MPTP (Otto et al., 1990). Tooyama et al. (1993) described a profound depletion of bFGF in surviving dopaminergic neurons in the substantia nigra of PD patients. Furthermore, bFGF was reported to have increased dopaminergic graft survival in a rat model of PD (Takayama et al., 1995).

Tumor necrosis factor-α (TNF-α) is a famous potent cytokine produced by activated macrophages. The concentration of TNF-α in the cerebrospinal fluid (CSF) was significantly higher in parkinsonian patients than those in controls (Mogi et al., 1995). Boka et al. (1995) detected TNF-immunoreactive glial cells in the substantia nigra of PD patients but not in their control subjects. As TNF-α is an early signal to mediate apoptosis, their observation may be an evidence to indicate the presence of apoptotic cell death in PD. TNF is known to promote glial reaction (Thery et al., 1992). In this respect, study of glial-neuronal interaction is an important subject to elucidate the degenerative process of PD.

Apoptosis could be a secondary phenomenon to mitochondrial respiratory failure and oxidative stress; they are the two major pathologenetic processes in PD (Mizuno et al., 1995; Schapira et al., 1990; Jenner et al., 1992).

As mentioned earlier, Dipasquale et al. (1991) reported apoptosis of culture cerebellar granular neurons by MPP$^+$; they used 50 μM of MPP$^+$; the question is whether or not this concentration inhibits mitochondrial respiration. In isolated mitochondria, 50 μM of MPP$^+$ induces approximately 50% inhibition of state 3 respiration (Mizuno et al., 1987). Suppose MPP$^+$ diffuses into granular neurons even in the absence of specific uptake site for MPP$^+$, this concentration is able to inhibit mitochondrial respiration. In our study (Mochizuki et al., 1994), 20 μM of MPP$^+$ induced apoptosis in ventral mesencephalic-striatal coculture. Hartley et al. (1994) and Itano et al. (1994) reported induction of apoptosis in PC12 cells by MPP$^+$. Rotenone, a specific inhibitor of complex I, also induced apoptosis (Hartley et al.,1994). Interestingly, low concentration (10 to 25 μM of MPP$^+$) caused apoptosis and high concentration necrosis (Hartley et al., 1994). Dopaminergic cells such as PC12 cells and cultured mesencephalic neurons have dopamine transporter through which MPP$^+$ is concentrated (Javitch et al., 1985). Therefore, in cultured cells, lower concentration could induce respiratory failure, and apoptosis observed in these experimental conditions may well be a result of mitochondrial respiratory failure. Furthermore, mitochondrial respiratory chain inhibitors other than complex I inhibitors such as antimycin A (complex III inhibitor) and oligomycin (complex V inhibitor) also can induce apoptosis in mammalian cells (Wolvetang et al., 1994).

Excitotoxicity has been widely studied and necrotic cell death has been the major type of excitotoxic cell death, i.e., cell swelling followed by membrane rupture (Choi et al., 1987; Siman and Card, 1988), but in a different condition, excitotoxins may induce apoptosis. Finiels et al. (1995) treated immature rat cortical cells cultured in the absence of serum with NMDA, glutamate, or quisqualate, and observed morphological evidence of apoptosis, i.e., cytoplasm and chromatin condensation; inhibitors of protein or RNA synthesis abolished cell death; but internucleosomal DNA degradation was not observed. Indirect excitotoxic injury to the striatum in developing rats (postnatal day 7) also induced apoptosis in the substantia nigra (Macaya et al., 1994); Macaya et al. (1994) injected quinolinate in developing rat striatum which induced decrease in dopaminergic neurons in the substantia nigra; quinolinate induced axon sparing injury; therefore, this was an indirect loss of nigral neurons; the morphologic characteristics of the dying cells were typical of apoptosis.

Role of oxidative stress in the induction of apoptosis has increasingly recognized in recent years. Kitajima et al. (1994) showed nitric oxide induced apoptosis in murine mastocytoma; inducible NO synthase (iNOS) mRNA was strongly expressed in apoptotic cells. Dopamine has been implicated as inducing oxidative stress as its oxidation by monoamine oxidase produces hydrogen peroxide. Ziv et al. (1994) reported apoptosis-like cell death in cultured chick sympathetic neurons induced by dopamine; they reported marked morphological alterations, mainly axonal disintegration and severe shrinkage and condensation of cell bodies. They postulated that this might be one of the mechanisms of nigral cell death in PD. Superoxide dismutase was reported to delay neuronal apoptosis (Greenlund et al., 1995); they used sympathetic neurons in culture which die when they were deprived of nerve growth factor; cultured sympathetic neurons were injected with Cu/Zn SOD protein or with an expression vector containing an SOD cDNA; in both cases apoptosis was delayed when the neurons were deprived of NGF; they postulate that superoxide production may occur early in response to trophic factor deprivation and that reactive oxygen species production serves as an early signal to mediate apoptosis.

Thus it appears to be likely that abnormal biochemical processes within nigral neurons induce apoptotic process.

Therapeutic implication if apoptosis is involved in neurodegeneration

If apoptosis is involved in neurodegeneration, neuroprotection might be achieved by inhibiting apoptosis. In PD, apoptosis appears to be located in the downstream of neurodegenerative processes; therefore, complete neuroprotection by inhibiting apoptosis appears to be difficult, however, at least partial protection may be possible. At lease, it is worthwhile to test such possibilities in experimental models first, and then in patients with PD.

In experimental conditions, numbers of substances are known to inhibit apoptosis effectively. For instance, sympathetic neurons die by apoptosis when deprived of nerve growth factor (NGF). Greenlund et al. (1995) injected superoxide dismutase protein or an expression vector containing an SOD cDNA; in both conditions, they observed delay in apoptosis of cultured sympathetic neurons which had been deprived of NGF; they postulated that reactive oxygen species would serve as an early signal to mediate apoptosis.

Bcl-2 was cloned as a novel gene present at the translocation breakpoint in follicular B-cell lymphoma by Tsujimoto et al. (1984); *bcl-2* is able to prevent apoptosis in numbers of conditions including cultured sympathetic neurons deprived of NGF (Garcia et al., 1992; Greenlund et al., 1995), PC12 cells deprived of glucose, serum and growth factors (Bredesen, 1995; Mah et al., 1993), glutamate toxicity in neural cell lines (Zhong et al., 1993), death of central neural cells induced by multiple agents such as calcium ionophore, free radical-inducing agents including *tert*-butyl hydroperoxide, glutamate, and β-amyloid peptide (Bredesen, 1995; Zhong et al., 1993), and cell death induced by reactive oxygen species (Kane et al, 1993; Myers et al., 1995). Over expression of *bcl-2* in transgenic mice protects neurons from naturally occurring cell death and experimental ischemia (Martinou et al., 1994). Expression of bcl-2 protein in neurons, glia cells and vascular cells was reported in human aged brain and neurodegenerative diseases (Migheli et al., 1994). However, there is a controversy regarding the expression of bcl-2 protein in neurons. In our hand, bcl-2 protein could not be detected in neurons in PD and AD, instead it was expressed in reactive microglia cells (Mochizuki et al., 1996, submitted).

The *p35* gene is expressed by the baculovirus Autographa californica nuclear polyhedrosis virus; expression of P35 protein prevents host cell apoptosis enabling marked reproduction of viral particles (Friesen and Miller, 1987). The *p35* gene is also expressed in mammalian neural cells inhibiting apoptosis induced by serum withdrawal, glucose withdrawal, and calcium ionophore (Rabizabeh et al., 1993). P35 protein prevents apoptosis by interleukin-1b-converting enzyme (ICE) family protease (Bump et al., 1995). ICE is a key enzyme that plays an important role in apoptotic cell death (Martin and Green, 1995). ICE protease has recently been purified and characterized, and named as apopain (Nicholson et al., 1995); they also found a potent peptide aldehyde inhibitor of ICE protease which could prevent apoptosis in vitro. But our knowledge on nervous system ICE is still limited. Recently, Tingsborg et al. (1996) reported the expression and distribution of ICE in the nervous system in the rat.

Glial cell line derived neurotrophic factor (GDNF) is a glycosylated disulfide-bounded homodimer that is a distantly related member of the transforming growth factor-β superfamily; GDNF specifically promotes the survival of dopaminergic neurons in rat embryonic cultures (Lin et al., 1993). Furthermore, Gash et al. (1996) described better functional recovery in motor performance of MPTP-treated monkeys by the treatment with GDNF. Development in the delivery system for such neurotrophic factors as GDNF may eventually enable the transport of those substances into the brain. Ex vivo transplantation using virus vector may be one of the useful methods of the delivery system for the neuroprotective agents into the brain (Takayama et al., 1995).

Many other genes that are upregulated or down regulated during apoptosis are know; they are either proapoptotic or anti-apoptotic (Bredesen, 1995). Therefore, by regulating expression of those genes, it may become possible to delay neuronal death in neurodegenerative disorders, and vast fields of research are open in this area.

Acknowledgements

This work was supported in part by Grant-in-Aid for Scientific Research on Priority Areas, Grant-in-Aid for Neuroscience Research from Ministry of Education, Science, and Culture, Japan, Grant-in-Aid for Neurodegenerative Disorders from Ministry of Health and Welfare, Japan, and "Center of Excellence" Grant from National Parkinson Foundation, Miami.

References

Anglade P, Vyas S, Jovoy-Agid F, Herrero MT, Michel PP, Marquez J, Mouatt-Prigent A, Ruberg M, Hirsch BC, Agid Y (1995) Apoptotic degeneration of nigral dopaminergic neurons in Parkinson's disease. Society for Neuroscience Abstract 21 (Part 1): 1250
Behl C, Davis B, Klier FG, Schubert H (1994) Amyloid beta peptide induces necrosis rather than apoptosis. Brain Res 645: 253–264

Boka G, Anglade P, Wallach D, Agid J, Agid Y, Hirsch EC (1994) Immunohistochemical analysis of tumor necrosis factor and its receptors in Parkinson's disease. Neurosci Lett 172: 151–154

Bredesen DE (1995) Neural apoptosis. Ann Neurol 38: 839–851

Brown RH, Jr (1995) Amyotrophic lateral sclerosis: recent insights form genetics and transgenic mice. Cell 80: 687–692

Bump NJ, Hackett M, Hugunin M, Seshagiri S, Brady K, Chen P, Ferenz C, Franklin S, Ghayur T, Li P, Licari P, Mankovich J, Shi L, Greenberg AH, Miller LK, Wong WW (1995) Inhibition of ICE family proteases by baculovirus antiapoptotic protein p35. Science 269: 1885–1888

Choi DW, Maulucci-Gedde MA, Kriegstein AR (1987) Glutamate neurotoxicity in cortical cell cultures. J Neurosci 7: 357–368

Clarke PG (1990) Developmental cell death: morphological diversity and multiple mechanisms. Anat Embryol 181: 195–213

Clem RJ, Miller LK (1994) Control of programmed cell death by the baculovirus genes p35 and IAP. Mol Cell Biol 14: 5212–5222

Cohen JJ (1991) Programmed cell death in the immune system. Adv Immunol 50: 55–85

Cohen JJ (1993) Apoptosis. Immunol Today 14: 126–130

Dipasquale B, Marini M, Youl RJ (1991) Apoptosis and DNA degradation by 1-methyl-4-phenylpyridinium in neurons. Biochem Biophys Res Commun 181: 1442–1448

Dragunow M, Faull RLM, Lawlor P, Beilharz EJ, Singleton K, Walker EB, Mee E (1995) In situ evidence for DNA fragmentation in Huntington's disease striatum and Alzheimer's disease temporal lobes. Neuro Report 6: 1053–1057

Engele J, Bohn MC (1991) The neurotrophic effects of fibroblast growth factors on dopaminergic neurons in vitro are mediated by mesencephalic glia. J Neurosci 11: 3070–3078

Esteban J, Rosen DR, Bowling AC, Sapp P, McKenna-Yasek D, O'Regan JP, Beal MF, Horvitz HR, Brown RH Jr (1994) Identification of two novel mutations and a new polymorphism in the gene for Cu/Zn superoxide dismutase in patients with amyotrophic lateral sclerosis. Hum Mol Genetics 3: 997–998

Forloni G, Chiesa R, Smiroldo S, Verga L, Salmone M, Tagliavini F, Angeretti N (1993) Apoptosis-/mediated neurotoxicity induced by chronic application of beta amyloid fragment. Neuro Report 4: 523–526

Finiels F, Robert JJ, Samolyk ML, Privat A, Mallet J, Revah F (1995) Induction of neuronal apoptosis by excitotoxins associated with long-lasting increase of 12-O-tetradecanoylphorbol 13-acetate-responsive element-binding activity. J Neurochem 65: 1027–1034

Friesen PD, Miller LK (1987) Divergent transcription of early 35- and 94-kilodalton protein genes encoded by the HindIIIK gene genome fragment of the baculovirus Autographa californica nuclear polyhedrosis virus. J Virol 61: 2264–2272

Gash DM, Zhang Z, Ovadia A, Cass WA, Yi A, Simmerman L, Russell D, Martin D, Lapchak PA, Collins F, Hoffer BJ, Gerhardt GA (1996) Functional recovery in parkinsonian monkeys treated with GDNF. Nature 380: 252–255

Garcia I, Martinou I, Tsujimoto, Martinou JC (1992) Prevention of programmed cell death of sympathetic neurons by the bcl-2 proto-oncogene. Science 258: 302–304

Gold R, Schmied M, Giegerich G, Breitschopf H, Hartung HP, Toyka KV, Lassmann H (1994) Differentiation between cellular apoptosis and necrosis by the combined use of in situ tailing and nick translation techniques. Lab Invest 71: 219–225

Grasl-Kraupp B, Ruttkay-Nedecky B, Koudelka H, Bukowska K, Bursch W, Schulte-Hermann R (1995) In situ detection of fragmented DNA (TUNEL assay) fails to discriminate among apoptosis, necrosis, and autolytic cell death: a cautionary note. Hepatology 21: 1465–1468

Greenlund LJS, Deckwerth TL, Johnson EM Jr (1995) Superoxide dismutase delays neuronal apoptosis: a role for reactive oxygen species in programmed neuronal death. Neuron 14: 303–315

Gurney ME, Pu H, Chiu AY, Dal Canto MC, Polchow CY, Alexander DD, Caliendo J, Hentati A, Kwon YW, Deng HX, Chen W, Zhai P, Sufit RL, Siddique T (1994) Motor neuron degeneration in mice that express a human Cu, Zn superoxide dismutase mutation. Science 264: 1772–1775

Hartley A, Stone JM, Heron C, cooper JM, Schapira AHV (1994) Complex I inhibitors induce dose-dependent apoptosis in PC12 cells: relevance to Parkinson's disease. J Neurochem 63: 1987–1990

Hengartner MP, Horvitz HR (1994) C. elegans cell survival gene ced-9 encodes a functional homolog of the mammalian protooncogene bcl-2. Cell 76: 665–676

Hiraishi K, Suzuki K, Hakomori S, Adachi M (1993) Le^Y antigen expression is correlated with apoptosis (programmed cell death). Glycobiology 3: 381–390

Itano Y, Kitamura Y, Nomura Y (1994) 1-Methyl-4-phenylpyridinium (MPP$^+$)-induced cell death in PC12 cells: inhibitory effects of several drugs. Neurochem Int 25: 419–425

Javitch JA, D'Amato RJ, Strittmatter SM, Snyder SH (1985) Parkinsonism-inducing neurotoxin, N-methyl-4-phenyl-1,2,3,6-tetrahydropyridine: uptake of the metabolite N-methyl-4-phenylpyridine by dopamine neurons explains selective toxicity. Proc Natl Acad Sci USA 82: 2173–2177

Jenner P, Dexter DT, Sian J, Schapira AHV, Marsden CG (1992) Oxidative stress as a cause of nigral cell death in Parkinson's disease and incidental Lewy body disease. Ann Neurol 32: S82–S87

Kane DJ, Sarafian TA, Anton R, Hahn H, Gralla EB, Valentine JS, Örd T, Bredesen DE (1993) Bcl-s inhibition of neural death: decreased generation of reactive oxygen species. Science 262: 1274–1277

Kerr JFR, Wyllie AH, Currie AR (1972) Apoptosis: a basic biological phenomenon with wide-ranging implications in tissue kinetics. Br J Cancer 26: 239–257

Kitajima I, Kawahara K, Nakajima T, Soejima Y, Matsuyama T, Maruyama I (1994) Nitric oxide-mediated apoptosis in murine mastocytoma. Biochem Biophys Res Commun 204: 244–251

Lassmann H, Bancher C, Breitschopf H, Wegiel J, Bobinski M, Jellinger K, Wisniewski HM (1995) Cell death in Alzheimer's disease evaluated by DNA fragmentation in situ. Acta Neuropathol 89: 35–41

Lin LF, Doherty D, Lile J, Bektesh S, Collins F (1993) GDNF: a glial cell line-derived neurotrophic factor for midbrain dopaminergic neurons. Science 260: 1130–1132

Liston P, Roy N, Tamai K, Lefebre C, Baird S, Cherton-Hovat G, Farahani R, McLean M, Ikeda JE, MacKenzie A, Korneluk RG (1996) Suppression of apoptosis in mammalian cells by NAIP and a related family of IAP genes. Nature 379: 349–353

Lockshin R, Beaulaton J (1974) Programmed cell death. Life Sci 15: 1549–1565

Lockshin RA, Williams CM (1964) Programmed cell death. II. Endocrine potentiation of the breakdown of the intersegmental muscles of silkmoths. J Insect Physiol 10: 643–649

Loo DT, Copani A, Pike CJ, Whitemore ER, Walencewicz AJ, Cotman CW (1993) Apoptosis is induced by β-amyloid in cultured central nervous system neurons. Proc Natl Acad Sci USA 90: 7951–7955

Macaya A, Munell F, Gubits RM, Burke RE (1994) Apoptosis in substantia nigra following developmental striatal excitotoxic injury. Proc Natl Acad Sci USA 91: 8117–8121

Mah SP, Zhong LT, Liu Y, Edwards RH, Bredesen DE (1993) The protooncogene bcl–2 inhibits apoptosis in PC12 cells. J Neurochem 60: 1183–1186

Martin SJ, Green DR (1995) Protease activation during apoptosis: death by a thousand cuts? Cell 82: 1–20

Martinou JC, Dubois-Dauphin M, Staple JK, Rodoriguez I, Frankowski H, Missotten M, Albertini P, Talabot D, Catsiċas S, Pietra C, Huarte J (1994) Overexpression of BCL-2 in transgenic mice protects neurons from naturally occurring cell death and experimental ischemia. Neuron 13: 1017–1030

Migheli A, Cavalla P, Piva R, Giordana MT, Schiffer D (1994) Bcl-2 protein expression in aged and neurodegenerative diseases. Clin Neurosci Neuropath 5: 1906–1908

Mizuno Y, Saitoh T, Sone N (1987) Inhibition of mitochondrial NADH-ubiquinone oxidoreductase activity by 1-methyl-4-phenylpyridinium ion. Biochem Biophys Res Commun 143: 294–299

Mizuno Y, lkebe S, Hattori N, Nakagawa-Hattori H, Mochizuki H, Tanaka M, Ozawa T (1995) Role of mitochondria in the etiology and pathogenesis of Parkinson's disease. Biochim Biophys Acta 1271: 265–274

Mochizuki H, Nakamura N, Nishi K, Mizuno Y (1994) Apoptosis is induced by 1-methyl-4-phenylpyridinium ion (MPP$^+$) in a ventral mesencephalic-striatal co-culture. Neurosci Lett 170: 191–194

Mochizuki H, Goto G, Mori H, Mizuno Y (1996) Histochemical detection of apoptosis in Parkinson's disease. J Neurol Sci (in press)

Mogi M, Harada M, Riederer P, Narabayashi H, Fujita K, Nagatsu T (1994) Tumor necrosis-α (TNF-α) increase both in the brain and in the cerebrospinal fluid from parkinsonian patients. Neurosci Lett 165: 208–210

Mogi M, Harada M, Kondo T, Riederer P, Inagaki H, Minami M, Nagatsu T (1994) Interleukin-1β, epidermal growth factor and transforming growth factor-α are elevated in the brain from parkinsonian patients. Neurosci Lett 180: 147–150

Munsat TL (1991) Workshop report: international SMA collaboration. Neuromusc Disord 1: 81

Myers KM, Fiskum G, Liu Y, Simmens SJ, Bredesen DE, Murphy AN (1995) Bcl-s protects neural cells from cyanide/aglycemia induced lipid oxidation, mitochondrial injury, and loss of viability. J Neurochem 65: 2432–2440

Nicholson DW, Ali A, Thornberry NA, Vaillancourt JP, Ding CK, Gallant M, Gareau Y, Griffin PR, Labelle M, Lazebnik YA, Munday NA, Raju SM, Smulson ME, Yamin TT, Yu VL, Miller DK (1995) Identification and inhibition of the ICE/CED-3 protease necessary for mammalian apoptosis. Nature 376: 37–43

Otto D, Unsicker K (1990) Basic FGF reverses chemical and morphological deficits in the nigrostriatal system of MPTP-treated mice. J Neurosci 10: 1912–1921

Portera-Cailliau C, Hedreen JC, Price DL, Koliatsos VE (1995) Evidence for apoptotic cell death in Huntington disease and excitotoxic animal models. J Neurosci 15: 3775–3787

Rabizabeh S, LaCount DJ, Friesen PD, Bredesen DE (1993) Expression of the baculovirus p35 gene inhibits mammalian neural cell death. J Neurochem 61: 2318–2321

Rabizabeh S, Gralla EB, Borchelt DR, Gwinn R, Valentine JS, Sisodia S, Wong P, Lee M, Hahn H, Bredesen DE (1995) Mutations associated with amyotrophic lateral sclerosis convert superoxide dismutase from antiapoptotic gene to a proapoptotic gene: studies in yeast and neural cells. Proc Natl Acad Sci USA 92: 3024–3028

Rosen ER, Siddique T, Patterson D, Figlewicz DA, Sapp P, Hentati A, Donaldson D, Goto J, O'Regan JP, Deng HX, Rahmani Z, Krizus A, McKenna-Yasek D, Cayabyab A, Gaston SM, Berger R, Tanzi RE, Halperin JJ, Herzfeldt B, Van den Bergh R, Hung WY, Bird T, Deng G, Mulder DW, Smyth C, Laing NG, Soriano E, Pericak-Vance MA, Haines J, Rouleau GA, Gusella S, Horvitz HR, Brown RH Jr (1993) Mutations in Cu/Zn superoxide dismutase gene are associated with familial amyotrophic lateral sclerosis. Nature 362: 59–62

Roy N, Mahadevan MS, McLean M, Shutler G, Yaraghi Z, Farahani R, Baird S, Besner-Johnston A, Lefebvre C, Kang X, Salih M, Aubry H, Tamai K, Guan X, Ioannou P, Crawford TO, de Jong PJ, Surh L, Ikeda J, Korneluk RG, MacKenzie A (1995) The gene for neuronal apoptosis inhibitory protein is partially deleted in individuals with spinal muscular atrophy. Cell 80: 167–178

Schapira AHV, Cooper JM, Dexter D, Clark JB, Jenner P, Marsden CD (1990) Mitochondrial Complex I deficiency in Parkinson's disease. J Neurochem 54: 823–827

Siman R, Card JP (1988) Excitatory amino acid neurotoxicity in hippocampal slice preparation. Neuroscience 26: 433–447

Su JH, Anderson A, Cummings BJ, Cotman CW (1994) Immunohistochemical evidence for apoptosis in Alzheimer's disease. Neuro Report 5: 2529–2533

Sulston JE, Horvitz HR (1977) Post-embryonic cell lineages of the nematode Caenorhabditis elegans. Dev Biol 82: 110–156

Takayama H, Ray J, Raymon HK, Baird A, Hogg J, Fisher LJ, Gage FH (1995) Basic fibroblast growth factor increases dopaminergic graft survival and function in a rat model of Parkinson's disease. Nature Medicine 1: 53–58

Thery C, Stanley ER, Mallat M (1992) Interleukin 1 and tumor necrosis factor-α stimulate the production of colony-stimulating factor 1 by murine astrocytes. J Neurochem 59: 1183–1186

Tingsborg S, Zetterstrom M, Alheim K, Hasanvan H, Schultzberg M, Bartfai T (1996) Regionally specific induction of ICE mRNA and enzyme activity in the rat brain and adrenal gland by LPS. Brain Res 712: 156–158

Tompkins MM, Hill WD (1995) Apoptotic-like changes in human substantia nigra. Society for Neuroscience Abstract 21 (Part 1): 1273

Tooyama I, Kawamata T, Walker D, Yamada T, Hanai K, Kimura H, Iwane M, Igarashi K, McGeer EG, McGeer PL (1993) Loss of basic fibroblast growth factor in substantia nigra neurons in Parkinson's disease. Neurology 43: 372–376

Tsujimoto Y, Finger L, Yunis J, Nowell PC, Croce CM (1984) Cloning of the chromosome breakpoint of neoplastic B cells with the t(14;18) chromosome translocation. Science 226: 1097–1099

Wolvetang EJ, Johnson KL, Krauer K, Ralph SJ, Linnaine W (1994) Mitochondrial respiratory chain inhibitors induce apoptosis. FEBS Lett 339: 40–44

Wyllie AH (1980) Glucocorticoid-induced thymocyte apoptosis is associated with endogenous endonuclease activation. Nature 284: 555–556

Yoshiyama Y, Yamada T, Asanuma K, Asahi T (1994) Apoptosis related antigen, Le^Y and nick-end labeling are positive in spinal motor neurons in amyotrophic lateral sclerosis. Acta Neuropathol 88: 207–211

Yuan J, Horvitz HR (1990) The Caenorhabditis elegans genes ced-3 and ced-4 act cell autonomously to cause programmed cell death. Dev Biol 138: 33

Yuan J, Shaham S, Ledoux S, Ellis HM, Horvitz HR (1993) The C. elegans cell death gene ced-3 encodes a protein similar to mammalian interleukin-1β-converting enzyme. Cell 75: 641–652

Ziv I, Melamed E, Nardi N, Luria D, Achiron A, Offen D, Barzilai A (1994) Dopamine induces apoptosis-like cell death in cultured chick sympathetic neurons – a possible novel pathogenetic mechanism in Parkinson's disease. Neurosci Lett 170: 136–140

Zhong LT, Kane D, Bredesen D (1993) BCL-2 blocks glutamate toxicity in neural cell lines. Mol Brain Res 19: 353–355

Zhong LT, Sarafian T, Kane DJ, Charles AC, Mah SP, Edwards RH, Bredesen DE (1993) bcl-s inhibits death of central neural cells induced by multiple agents. Proc Natl Acad Sci USA 90: 4533–4537

Authors' address: Dr. Y. Mizuno, Department of Neurology, Juntendo University School of Medicine, 2-1-1 Hongo, Bunkyo, Tokyo 113, Japan.

Mechanisms of cell death in Alzheimer's disease

C. Bancher[1], **H. Lassmann**[2], **H. Breitschopf**[2], and **K.A. Jellinger**[1]

[1]Ludwig Boltzmann Institute of Clinical Neurobiology, Department of Neurology, Lainz Hospital, Vienna, [2]Neurological Institute, University of Vienna, Vienna, Austria

Summary. The etiology of Alzheimer's disease (AD) as well as its exact pathogenesis are unknown. Eventhough the deposition of βA4 and the formation of neurofibrillary tangles represent impressive morphological hallmarks of the disease, several lines of evidence suggest that both lesions are not sufficient as causes of the neurodegenerative process. On the other hand, in vitro studies have shown that βA4 is neurotoxic and is able to induce apoptotic cell death in neuronal cell cultures. Cells dying by apoptosis (programmed cell death) can be visualized in the tissue with a molecular biologic technique detecting fragmented nuclear DNA. Using this method, we have detected 50 × more neurons and 25 × more glial cells with nuclear DNA fragmentation in the brains of patients with AD than in non-demented controls. In contrast to previous studies, most of these cells did not reveal the characteristic morphological hallmarks of apoptosis. Most dying cells were not located within amyloid deposits and most dying cells did not bear a tangle. On the other hand, being in physical contact with an amyloid deposit increased the risk of a cell to dye by factor 5.7 and carrying a neurofibrillary tangle imposed a 3 times higher risk compared to unaffected nerve cells. Taken together, these data indicate that nerve cell death in AD occurs via a mechanism of programmed cell death different from classical apoptosis. Eventhough plaques and tangles increase the risk of cells to degenerate, both lesions are not the sole responsibles of the degenerative process, suggesting the existence of other factors that trigger the initiation of the cell death program in AD.

Introduction

The cause and the mechanism by which neurons die in Alzheimer's disease (AD) are unknown. From the structural changes observed in AD brains, the deposition of amyloid in senile plaques (SP) and the formation of neurofibrillary tangles (NFT) in neurons represent impressive morphological hallmarks of the disease and have long been considered sufficient as a cause of the neurodegenerative process. Indeed, from the histopathologic picture of an AD brain, it is easily conceivable that tangle-bearing neurons do not function and do not remain viable. It also appeared likely that the deposition of large amounts of amyloid surrounded by dystrophic neurites in the neuropil leads to disruption of the neuronal network, disconnection of intracortical pathways and clinically evident dementia. Early clinicopathologic studies have supported this view (Blessed et al., 1968; Tomlinson and Henderson 1976; Wilcock and Esiri 1982).

Do tangles kill neurons?

More recently, however, several lines of evidence have challenged the pathogenetic relevance of the classical structural lesions. For the case of NFT, studies on patients with Down's syndrome (Mann et al., 1986; Mann and Esiri 1989) and a number of clinicopathological correlation studies in AD have suggested that these lesions, eventhough the best histopathologic correlate of severe dementia, only occur at a late stage of progression of the disease (Fischer et al., 1991; Delaere et al., 1989; Wettstein et al., 1990; Jellinger et al., 1992; Bancher et al., 1996a). In addition, NFT are also found in a number of unrelated central nervous system diseases (Wisniewski et al., 1979; Bancher et al., 1996b), some of which have known etiologies (viral, traumatic, toxic, ...) and in which the formation of NFT occurs as an epiphenomenon. The formation of NFT is thus an unlikely candidate as a primary cause of dementia.

βA4 is neurotoxic and can induce apoptosis in tissue culture

The pathogenetic relevance SP has been supported by studies that have examined the effect of βA4 on neuronal cell cultures. Preliminary data have yielded conflicting results, since βA4 was shown to have both, neurotrophic and neurotoxic effects (Yankner et al., 1990, 1991), with GABAergic neurons being resistant against βA4 induced toxicity (Pike and Cotman 1993). This issue was resolved as it turned out that the physical aggregation state of the peptide had an influence on its activity on tissue culture cells: polymerization of the peptide into amyloid fibrils was necessary to produce neurotoxicity (Pike et al., 1993). In vitro studies that have adressed the mode of cell death occurring in βA4 treated cultures have suggested that death of the cells occurs either via apoptotic (Loo et al., 1993; Forloni et al., 1993a; Le et al., 1995) or necrotic mechanisms (Behl et al., 1992). More recently, it was suggested that the form of βA4 induced cell death (apoptosis or necrosis) is dependent on the cell type (Gschwind and Huber 1995).

Apoptosis

Apoptosis is a mode of programmed cell death that occurs physiologically during brain development and plays a major role in the regulation of inflammatory processes. In contrast to necrosis, apoptotic cell death is gene directed and represents an active phenomenon of the cell. By morphology and biochemistry, apoptosis is characterized by a number of features that differentiate it from other modes of cellular death. Characteristic features of apoptosis include early condensation and margination of chromatin at the periphery of the nucleus, plasma membrane blebbing giving rise to so-called apoptotic bodies, long lasting integrity of the plasma membrane, and internucleosomal fragmentation of nuclear DNA resulting in DNA pieces with multiples of a single nucleosome length, which is approximately 180 base pairs. This results in the formation of the typical "ladder" when DNA of apoptotic cells is run on electrophoretic gels (Lo et al., 1995; Bredesen 1995). The free ends of DNA fragments are the targets of the enzyme terminal transferase which can be used to visualize apoptosis in cells. We hypothesized that if neurons in AD dye by apoptosis, then it should be possible to detect these cells in brain tissue of AD patients. If βA4 has neurotoxic effects in vivo and induces apoptosis in the AD brain, then there should be some relationship of dying cells to βA4 deposits.

Material and methods

Ways to detect dying cells in AD brain tissue

Even though morphologic criteria are the most consistent determinants of apoptosis, molecular biologic techniques allowing the detection of DNA strand breaks can aid in identifying the cells that undergo apoptosis. A widely used method utilizes the enzymes DNA polymerase or terminal transferase to incorporate labelled oligonucleotides into the fragmented DNA of apoptotic cells (TUNEL method, in situ tailing) (Iseki, 1986; Gavrieli et al., 1992) and allows the differentiation of apoptotic from necrotic cell death (Gold et al., 1994).

We have used a modification of the TUNEL method (Gold et al., 1993, 1994) to test whether apoptotic cells or cells displaying DNA fragmentation can be detected in the brains of 18 AD patients obtained between 5.5 and 45h after death and 15 age- and post mortem time matched controls. Labelled slides were immunohistochemically counterstained with antibodies to βA4 (4G8, Kim et al., 1988) to detect amyloid deposits, PHF-ubiquitin (3–39, Wang et al., 1984) to detect tangle-bearing neurons, GFAP for astrocytes, myelin oligodendroglia glycoprotein (8–18/C5) for oligodendroglial cells and ferritin to visualize microglia. Labeled cells were counted in the temporal isocortex and in various allocortical areas of the inferior temporal lobe.

Results

Cells with nuclear DNA fragmentation in AD brains: an artefact?

Nuclei with DNA fragmentation were detected in all cases, the number of labeled cells being significantly higher in the AD cases than in controls. At first, we were concerned about the possibility that tissue preparation or post mortem DNA degradation may be responsible for the DNA strand breaks detected. If a postmortal artefact was the reason for positive reaction, then increasing post mortem time should go parallel with increasing numbers of

144   C. Bancher et al.

labeled cells. There was, however, no correlation between the duration of the
post mortem interval and the number of labeled cells (Fig. 1). In addition, in
a rat model of autoimmune encephalitis where high numbers of apoptotic
lymphocytes are present, there was no increased number of labeled cells with
post mortem time up to 24 hours. A recent methodological study on rat and
human brain has confirmed this result (Lucassen et al., 1995). The DNA
fragmentation observed in AD brains is thus not due to postmortal DNA
degradation but represents a true intravital phenomenon.

Which cells degenerate in AD?

Labeled cells were not only neurons, but also large numbers of oligodendroglial and microglial cells. On average, there were 50 times more neurons and
25 times more glial cells labeled in AD than in controls (Table 1). Evaluation
of immunohistochemically counterstained sections confirmed this result. In
AD brains, the largest numbers of degenerating cells were found in the
temporal allocortex. In these brains, an average of 28% of all labeled cells
could be identified as neurons. Other degenerating cells were mostly oligodendrocytes, preferentially seen in the subcortical white matter underlying
areas with severe neuronal loss, and microglia that was frequently seen in
association with SP. Most labeled nuclei did not reveal the typical morphological features of apoptosis (condensed, marginated chromatin). The cytoplasm
of these cells was frequently swollen or appeared empty around the labeled
nucleus. In the control brains, the vast majority of dying cells were located in
the white matter and could be identified as oligodendrocytes. Degenerating
neurons or astrocytes were extremely rare in these cases (Table 1).

Is there a relationship between dying cells, plaques and tangles?

Analysis of sections immunocytochemically counterstained with the antibody
to βA4 revealed that only 27% (range: 13–50%) of the degenerating cells were
located within areas of neuropil covered by amyloid. However, of the popula-

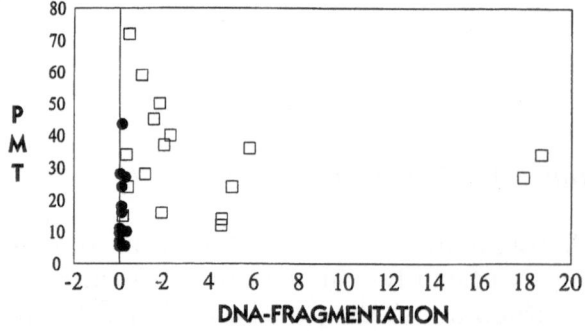

Fig. 1. Relationship between post mortem delay and the number of cells with positive
staining for DNA fragmentation. Black circles: AD cases (n = 18). Open boxes: Controls
(n = 15). Reprinted with permission from Lassmann et al., 1995

tion of cells located within amyloid deposits, 5.7 × (± 0.8) more cells displayed DNA fragmentation than in the population of cells located in unaffected tissue. This result indicates that, (eventhough most degenerating cells are located outside amyloid deposits), cell bodies with direct physical contact to βA4 have a higher risk to degenerate than those without. Similarly, evaluation of slides counterstained with anti-PHF/ubiquitin revealed that only 41% (range: 18–66%) of degenerating neurons carry a NFT; however, these neurons are at a 3 × (± 0.5) higher risk to degenerate than those devoid of a NFT (Lassmann et al., 1995) (Fig. 2).

Table 1. Numbers of cells with DNA fragmentation in the brains of AD patients and controls

	Alzheimer's disease	Controls
Number of patients	18	15
Mean age (yrs)	78 ± 2	73 ± 4
Disease duration (yrs)	6.4 ± 3.7	–
Amyloid deposition (% area) temporal lobe	6.8 ± 0.8*	0.7 ± 0.3
Tangles temporal lobe	13.3 ± 2.4*	1.1 ± 0.5
DNA-fragmentation total labeled cells	3.9 ± 1.3*	0.14 ± 0.4
Labeled neurons	1.1 ± 0.4*	0.02 ± 0.01
Labeled glia	2.8 ± 1.0*	0.12 ± 0.03

* p = 0.0001; values given in labeled cells/mm^2. Reprinted with permission from Lassmann et al., 1995

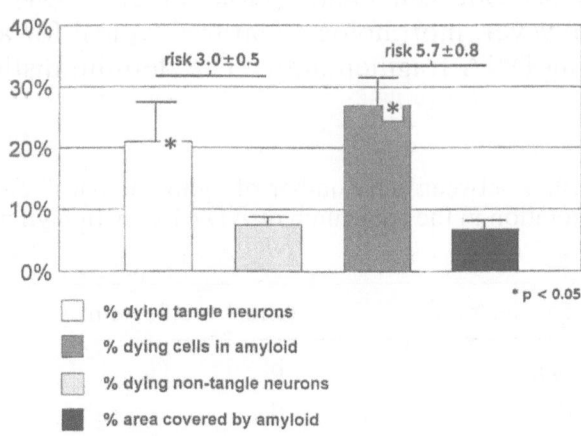

Fig. 2. Percentage of cells displaying DNA fragmentation in relationship to amyloid deposits and NFT. Reprinted with permission from Lassmann et al., 1995. The increased risk of neurons to die when bearing a tangle is calculated by dividing the percentage of degenerating cells in the tangle-bearing neuron population by the percentage of degenerating cells in the tangle-free population. The increased risk of cells to die when located within an amyloid deposit is determined by dividing the percentage of degenerating cells within amyloid deposits by the percentage of brain tissue covered by amyloid deposits

If the number of cells with DNA fragmentation is a measure of the acuity of the degenerating process, then it should be expected that remnants of dying cells be more frequent in cases with high numbers of labeled nuclei. Remnants of degenerating cells can usually not be found in AD brain tissue, but those degenerating cells that carry NFT are said to leave their NFT free in the neuropil as so-called "ghost tangles" or "tombstones". Indeed, an analysis of the subiculum of six cases with high numbers of ghost tangles revealed that large numbers of labeled neurons were associated with high numbers of ghost tangles in the same area. Labelling of the ghost tangles themselves was exceptional and was probably due to remnants of the disintegrated nucleus trapped between PHF (Table 2).

Discussion

Apoptosis, necrosis, or other form of cell death?

We have detected cells with nuclear DNA fragmentation in the brains of patients with AD. Not only neurons, but also high numbers of oligodendrocytes and microglial cells had nuclei with fragmented DNA. Integrity of nuclear DNA is a prerequisite for the survival of cells, thus, the cells detected are those undergoing degeneration.

A few histochemical studies using the TUNEL method have suggested that neurons (Su et al., 1994; Cotman and Anderson, 1995; Dragunow et al., 1995) and astrocytes (Smale et al., 1995) undergo apoptosis in AD brains. The assumption that the death of these cells occurs via an apoptotic mechanism is mainly based on positive reaction with the TUNEL method. Indeed, the use of in situ tailing with terminal transferase has been suggested to discriminate apoptotic from necrotic cell death (Gold et al., 1994). According to our observations, however, morphologic features typical for apoptosis were rare in cells displaying DNA fragmentation. The definite distinction of apoptosis

Table 2. Relationship between the number of "ghost tangles" and the number of cells with DNA fragmentation in the subiculum of AD cases with high numbers of allocortical NFT

Case	Tangles/mm^2	Ghost tangles/mm^2	DNA-Frag/mm^2
A	9.6	0	3.3
B	18.2	0	3.4
C	21.0	10	6.4
D	54.0	11	23.5
E	33.0	17	41.0
F	40.0	59	52.5

DNA-Frag Neurons with nuclear DNA fragmentation. Reprinted with permission from Lassmann et al., 1995

from necrosis or other forms of cell death requiring both, positive reaction with in situ tailing and morphological description, the absence of typical cellular changes indicates that cell death in AD proceeds by another form of programmed cell death than classical apopotosis.

Apoptosis in the central nervous system

In addition to central nervous system (CNS) development, apoptotic cell death has been suggested to occur in a variety of pathological conditions of the CNS using biochemical and in situ techniques. Among them are neurodegenerative diseases such as Parkinson's disease (Agid, 1995; Anglade et al., 1996), amyotrophic lateral sclerosis (Yoshiyama et al., 1994) and Huntingtons's disease (Dragunow et al., 1995; Portera-Cailliau et al., 1995; Thomas et al., 1995). TUNEL positive cells have also been observed in glioblastoma multiforme and traumatic brain injury (Thomas et al., 1995). In HIV encephalitis, apopotic cell death has been documented by both, biochemical as well as histochemical methods (Adlebiassette et al., 1995), and the dying cells have been indentified as neurons and astrocytes (Petito and Roberts, 1995). In vitro studies have demonstrated that, similar to βA4, a fragment of the prion protein has the ability to induce apoptotic cell death and the expression of c-Fos in rat hippocampal cultures (Forloni et al., 1993b). Apoptotic cell death has also been shown to occur in experimental settings such as in kainic acid induced epilepsy (Pollard et al., 1994), retrogradely degenerating central cholinergic medial septal nucleus neurons after transsection of the fimbria fornicis (Wilcox et al., 1995) and excitotoxic injury of rat striatum, where apoptosis and necrosis have been shown to occur simultaneously (Portera-Cailliau et al., 1995). In this context, it is interesting to note that apoptosis and necrosis are not exclusive mechanisms of cell death and that the intensity of the stimulus appears to be a decisive factor in whether cells undergo apoptotic or necrotic death. Indeed, exposure of cortical neurons to short durations or low concentrations of the excitotoxin NMDA induces a delayed form of cell death predominated by apoptotic features whereas intense exposure to high concentrations of NMDA leads to necrotic cell damage (Bonfoco et al., 1995a). In addition, primarily apoptotic cell death/damage may be followed by secondary necrosis (Majno and Joris, 1995).

Apoptosis-associated proteins play a role in cell death in AD

Apoptosis has been shown to be associated with the expression of a number of so-called apoptosis-associated proteins, some of which belong to the Jun- and Fos-family of polypeptides (Anderson et al., 1994). E.g., cerebellar granule cells exposed to colchicine, a microtubule-disrupting agent, undergo apoptosis and express c-Fos, indicating that cytoskeletal alterations can initiate programmed cell death (Bonfoco et al., 1995b). A recent study has demonstrated that treatment of cultured hippocampal neurons with βA4 can induce the expression of c-Jun in vulnerable, but not in resistant cells, and that the cells expressing c-Jun display changes in nuclear morphology indicative of apoptosis (Anderson et al., 1995). Another apoptosis-associated protein, Fas-antigen (CD 95) has been shown to be expressed in senile plaques and a subset of reactive astrocytes in AD brain tissue (Nishimura et al., 1995). On the other hand, the apoptosis-preventing protein Bcl-2 is also elevated in AD brain tissue and is expressed in astrocytes (Obarr et al., 1996) and in hippocampal and entorhinal neurons, whereby Bcl-2 appears to be downregulated in neurons carrying NFT. Bcl-2 may have a role in compensation responses to AD pathology, perhaps affording to the remaining neurons a margin of protection from apoptosis (Satou et al., 1995). These data taken toghether, and in particular the fact that apoptosis-related antigens are up-regulated and Bcl-2 is reduced in cells undergoing degeneration mediated by AD pathology corroborate the notion the apoptosis or other gene-directed ways of cell death play a role in the AD neurodegenerative process.

Mechanism of βA4 mediated cell death

The mechanism of death of cells undergoing βA4 neurotoxicity-mediated apopotosis has been addressed in a number of studies. The apoptosis-mediated neuronal cell death program appears to reside in the ability of βA4 to activate NMDA-gated Ca^{2+} channels in neuronal membranes resulting in an excessive Ca^{2+} influx and an amount of formation of free radicals that exceeds the defensive capacity of cells (Le et al., 1995). Indeed, the neurotoxic action of βA4 can be inhibited by certain NMDA agonists and antagonists (Le et al., 1995; Copani et al., 1991, 1995). In addition, βA4 has the ability to accelerate neuronal degeneration in the presence of defective glucose metabolism (Copani et al., 1991), and to potentiate the toxicity of excitotoxins such as glutamate, NMDA and kainate (Koh et al., 1990), whose cytotoxic action can in turn enhance that of βA4 (Le et al., 1995). NMDA-mediated neurotoxicity appears thus to play a role in the βA4-induced programmed cell death. These facts are of interest in relation to our finding that being in physical contact with βA4 deposits imposes an elevated risk to the affected neuron to degenerate, without being lethal by itself. This underscores the notion that the impact of amyloid deposition needs additional pathogenic factors to induce nerve cell death.

Too many degenerating neurons for a slowly progressive disease

In tissue culture, apoptotic cell death takes only a few hours (Kerr et al., 1972). On the other hand, the density of nuclei with DNA fragmentation was high in the subiculum of some of the AD cases (up to 52/mm²). This incidence of degenerating cells would correspond to a total depletion of subicular neurons within a few days and is much higher than what can be expected in a chronic, slowly progressive disease. On the other hand, some AD cases with equally severe AD-type neuropathology did not reveal such extensive cell degeneration, suggesting that the ultimate cell death in AD requires other precipitating factors. The rate of acute cell death may increase in the preterminal phase of the illness, probably due to synergy of the disease-specific process with other factors, such as hypoxia and vascular or metabolic disturbances that are more likely to occur at the terminal stage (Lassmann et al., 1995).

Conclusions

In conclusion, our findings suggest that at least some neuronal death in AD occurs via a mechanism of programmed cell death different from classical apoptosis. Both neurons and glia appear to be at a higher risk of degeneration in AD than in control brains, whereby some of this risk is imposed by neurotoxic effects of βA4 and the accumulation of PHF within neurons. Both lesions, however, are not the sole responsibles of the AD neurodegenerative process, suggesting a multifactorial pathogenesis of dementia. Further studies will be necessary to identify other factors that trigger the initiation of the cell death program in neurodegenerative diseases.

Acknowledgements

The authors wish to thank Ms. E. Gurnhofer for expert technical assistance. This project was supported by the Austrian Federal Ministry of Science, Research and Arts.

References

Adlebiassette H, Levy Y, Colombel M, Poron F, Natchev F, Keohane C, Gray F (1995) Neuronal apoptosis in HIV infection in adults. Neuropathol Appl Neurobiol 21: 218–227

Agid Y (1995) Aging, disease and nerve cell death. J Neural Transm [GenSect] 102: I (abstr)

Anderson AJ, Cummings BJ, Cotman CW (1994) Increased immunoreactivity for Jun- and Fos-related proteins in Alzheimer's disease: Association with pathology. Exp Neurol 125: 286–295

Anderson AJ, Pike CJ, Cotman CW (1995) Differential induction of immediate early gene proteins in cultured neurons by β-amyloid (A-β) Association of c-Jun with A-β induced apoptosis. J Neurochem 65: 1487–1498

Anglade P, Vyas S, Javoy-Agid F, Herrero MT, Michel PP, Marquez J, Mouatt-Prigent A, Ruberg M, Hirsch EC, Agid Y (1996) Apoptotic degeneration of nigral dopaminergic neurons in Parkinson's disease. Neurology 46: A467 (abstr)

Bancher C, Jellinger K, Lassmann H, Fischer P, Leblhuber F (1996a) Correlations between mental state and quantitative neuropathology in the Vienna Prospective Longitudinal Study on Dementia. Eur Arch Psychiatry Clin Neurosci 246: 137–146

Bancher C, Leitner H, Jellinger K, Eder H, Setinek U, Fischer P, Wegiel J, Wisniewski HM (1996b) On the relationship between measles virus and Alzheimer neurofibrillary tangles in subacute sclerosing panencephalitis. Neurobiol Aging 17: 527–533

Behl C, Davis JB, Klier FG, Schubert D (1994) Amyloid β peptide induces necrosis rather than apoptosis. Brain Res 645: 253–264

Blessed G, Tomlinson BE, Roth M (1968) The association between quantitative measures of dementia and of senile change in the cerebral gray matter of elderly subjects. Brit J Psych 114: 797–811

Bonfoco E, Krainc D, Ankarcrona M, Nicotera P, Lipton SA (1995a) Apoptosis and necrosis – two distinct events induced, respectively, by mild and intense insults with N-methyl-D-aspartate or nitric oxide/superoxide in cortical cell cultures. Proc Natl Acad Sci USA 92: 7162–7166

Bonfoco E, Cecatelli S, Manzo L, Nicotera P (1995b) Colchicine induces apoptosis in cerebellar granule cells. Exp Cell Res 218: 189–200

Bredesen DE (1995) Neural apoptosis. Ann Neurol 38: 839–851

Copani A, Koh J-Y, Cotman CW (1991) β-amyloid increases neuronal susceptibility to injury by glucose deprivation. Neuro Report 2: 763–765

Copani A, Bruno V, Battaglia G, Leanza G, Pellitteri R, Russo A, Stanzani S, Nicoletti F (1995) Activation of metabotropic glutamate receptors protects cultured neurons against apoptosis induced by β-amyloid peptide. Mol Pharmacol 47: 890–897

Cotman CW, Anderson AJ (1995) A potential role for apoptosis in neurodegeneration and Alzheimer's disease. Mol Neurobiol 10: 19–45

Crystal H, Dickson D, Fuld P, Masur D, Scott R, Mehler M, Masdeu J, Kawas C, Aronson M, Wolfson L (1988) Clinico-pathological studies in dementia: Nondemented subjects with pathologically confirmed Alzheimer's disease. Neurology 38: 1682–1687

Delaere P, Duyckaerts C, Brion JP, Poulain V, Hauw JJ (1989) Tau, paired helical filaments and amyloid in the neocortex: a morphometric study of 15 cases with graded intellectual status in aging and senile dementia of Alzheimer's type. Acta Neuropathol 77: 645–653

Dickson DW, Crystal HA, Mattiace LA, Masur DM, Blau AD, Davies P, Yen S-H, Aronson MK (1991) Identification of normal and pathologic aging in prospectively studied nondemented elderly humans. Neurobiol Aging 13: 179–189

Dragunow M, Faull RLM, Lawlor P, Beilharz EJ, Singleton K, Walker EB, Mee E (1995) In situ evidence for DNA fragmentation in Huntigton's disease striatum and Alzheimer's disease temporal lobes. Neuro Report 6: 1053–1057

Fischer P, Lassmann H, Jellinger K, Simanyi M, Bancher C, Travniczek-Marterer A, Gatterer G, Danielczyk W (1991) Die Demenz vom Alzheimer Typ. Eine klinische Längsschnittstudie mit quantitativer Neuropathologie. Wien Med Wochenschr 141: 455–462

Forloni G, Chiesa R, Smiroldo S, Verga L, Salmona M, Tagliavini F, Angeretti N (1993a) Apoptosis mediated neurotoxicity induced by chronic application of β-amyloid fragment 25–35. Neuro Report 4: 523–526

Forloni G, Angeretti N, Chiesa R, Monzani E, Salmona M, Bugiani O, Tagliavini F (1993b) Neurotoxicity of a prion protein fragment. Nature 362: 543–546

Gavrieli Y, Sherman Y, Ben-Sasson SA (1992) Identification of programmed cell death in situ via specific labeling of nuclear DNA fragmentation. J Cell Biol 119: 493–501

Gold R, Schmied M, Rothe G, Zischler H, Breitschopf H, Wekerle H, Lassmann H (1993) Detections of DNA fragmentation in apoptosis: application of in situ nick translation to cell culture systems and tissue sections. J Histochem Cytochem 41: 1023–1030

Gold R, Schmied M, Giegerich G, Breitschopf H, Hartung HP, Toyka K, Lassmann H (1994) Differentiation between cellular apoptosis and necrosis by the combined use of in situ tailing and nick translation techniques. Lab Invest 71: 219–225

Gschwind M, Huber G (1995) Apoptotic cell death induced by β-amyloid 1–42 peptide is cell type dependent. J Neurochem 65: 292–300

Iseki S (1986) DNA strand breaks in rat tissue as detected by in situ nick translation. Exp Cell Res 167: 311–326

Jellinger K, Bancher C, Fischer P, Lassmann H (1992) Quantitative histopathologic validation of senile dementia of the Alzheimer type. Eur J Gerontol 3: 146–156

Katzman R, Terry R, DeTeresa R, Brown T, Davies P, Fuld P, Renbing X, Peck A (1988) Clinical, pathological, and neurochemical changes in dementia: A subgroup with preserved mental status and numerous neocortical plaques. Ann Neurol 23: 138–144

Kerr JFR, Wyllie AH, Currie AR (1972) Apoptosis: a basic biological phenomenon with wide-ranging implications in tissue kinetics. Br J Cancer 26: 239–257

Kim KS, Miller DL, Sapienza VJ, Chen CM, Bai C, Grundke-Iqbal I, Currie JR, Wisniewski HM (1988) Production and characterization of monoclonal antibodies reactive to synthetic cerebrovascular amyloid peptide. Neurosci Res Commun 2: 121–130

Koh J-Y, Yang LL, Cotman CW (1990) β-amyloid protein increases the vulnerability of cultured cortical neurons to excitotoxic damage. Brain Res 533: 315–320

Lassmann H, Bancher C, Breitschopf H, Wegiel J, Bobinski M, Jellinger K, Wisniewski HM (1995) Cell death in Alzheimer's disease evaluated by DNA fragmentation in situ. Acta Neuropathol 89: 35–41

Le WD, Colom LV, Xie WJ, Smith RG, Alexianu M, Appel SH (1995) Cell death induced by β-amyloid 1–40 in MES 23.5 hybrid clone – the role of nitric oxide and NMDA-gated channel activation leading to apoptosis. Brain Res 686: 49–60

Lo AC, Houenou LJ, Oppenheim RW (1995) Apoptosis in the nervous system – Morphological features, methods, pathology, and prevention. Arch Histol Cytol 58: 139–149

Lockhart BP, Benicourt C, Junien JL, Privat A (1994) Inhibitors of free radical formation fail to attenuate direct β-amyloid (23–35) peptide-mediated neurotoxicity in rat hippocampal cultures. J Neurosci Res 39: 494–505

Loo DT, Copani A, Pike CJ, Whittemore ER, Walencewicz AJ, Cotman CW (1993) Apoptosis is induced by β-amyloid in cultured central nervous system neurons. Proc Natl Acad Sci USA 90: 7951–7955

Lucassen PJ, Chung WCJ, Vermeulen JP, Vanlookeren M, Vandierendonck CJH, Swaab DF (1995) Microvawe-enhanced in situ end-labeling of fragmented DNA – parametric studies in relation to postmortem delay and fixation of rat and human brain. J Histochem Cytochem 43: 1163–1171

Majno G, Joris I (1995) Apoptosis, oncosis and necrosis. An overview of cell death. Am J Pathol 146: 3–15

Mann DMA, Yates PO, Marcyniuk B, Ravindra CR (1986) The topography of plaques and tangles in Down's syndrome patients of different ages. Neuropathol Appl Neurobiol 12: 447–457

Mann DMA, Esiri M (1989) The pattern of acquisition of plaques and tangles in the brains of patients under 50 years of age with Down's syndrome. J Neurol Sci 89: 169–179

Masliah E, Terry RD, Mallory M, Alford M, Hansen LA (1990) Diffuse plaques do not accentuate synapse loss in Alzheimer's disease. Am J Pathol 137: 1293–1297

Morris JC, McKeel DW Jr, Storandt M, Rubin EH, Price JL, Grant EA, Ball MJ, Berg L (1991) Very mild Alzheimer's disease: Informant-based clinical, psychometric, and pathologic distinction from normal aging. Neurology 41: 469–478

Morris JC, Storandt M, McKeel DW Jr, Rubin EH, Price JL, Grant EA, Berg L (1996) Cerebral amyloid deposition and diffuse plaques in "normal" aging: Evidence for presymptomatic and very mild Alzheimer's disease. Neurology 46: 707–719

Nishimura T, Akiyama H, Yonehara S, Kondo H, Ikeda K, Kato M, Iseki E, Kosaka K (1995) Fas antigen expression in brains of patients with Alzheimer-type dementia. Brain Res 695: 137–145

Obarr S, Schultz J, Rogers J (1996) Expression of the protooncogene Bcl-2 in Alzheimer's disease brain. Neurobiol Aging 17: 131–136

Petito CK, Roberts B (1995) Evidence of apoptotic cell death in HIV encephalitis. Am J Pathol 146: 1121–1130

Pike CJ, Cotman CW (1993) Cultured GABA-immunoreactive neurons are resistant to toxicity induced by β-amyloid. Neuroscience 56: 269–274

Pike CJ, Burdick D, Walencewicz AJ, Glabe CG, Cotman CW (1993) Neurodegeneration induced by β-amyloid peptides in vitro: The role of peptide assembly state. J Neurosci 13: 1676–1687

Pollard H, Cantagrel S, Charriaut-Marlangue C, Moreau J, Ari YB (1994) Apoptosis associated DNA fragmentation in epileptic brain damage. Neuro Report 5: 1053–1055

Portera-Cailliau C, Hedreen JC, Price DL, Koliatsos VE (1995) Evidence for apoptotic cell death in Huntington disease and excitotoxic animal models. J Neurosci 15: 3775–3787

Satou T, Cummings BJ, Cotman CW (1995) Immunoreactivity for Bcl-2 protein within neurons in the Alzheimer's disease brain increases with disease severity. Brain Res 697: 35–43

Selkoe DJ (1994) Alzheimer's disease: a central role for amyloid. J Neuropathol Exp Neurol 53: 438–447

Smale G, Nichols NR, Brady DR, Finch CE, Horton WE (1995) Evidence for apoptotic cell death in Alzheimer's disease. Exp Neurol 133: 225–230

Su JH, Anderson AJ, Cummings BJ, Cotman CW (1994) Immunohistochemical evidence for apoptosis in Alzheimer's disease. Neuro Report 5: 2529–2533

Thomas LB, Gates DJ, Richfield EK, O'Brien TF, Schweitzer JB, Steindler DA (1995) DNA end labeling (TUNEL) in Huntington's disease and other neuropathological conditions. Exp Neurol 133: 265–272

Tomlinson BE, Henderson G (1976) Some quantitative cerebral findings in normal and demented old people. In: Terry RD, Gershon S (eds) Neurobiology of Aging. Raven Press, New York, pp 183–204

Wang GP, Grundke-Iqbal I, Kascak RJ, Iqbal K, Wisniewski HM (1984) Alzheimer neurofibrillary tangles: monoclonal antibodies to inherent antigens. Acta Neuropathol 62: 268–275

Wettstein A, Lang W (1990) Correlation of cognitive skills in nursing home patients with histologic Alzheimer changes in specific brain areas. Dementia 1: 278–285

Wilcock GK, Esiri MM (1982) Plaques, tangles and dementia. A quantitative study. J Neurol Sci 56: 343–356

Wilcox BJ, Applegate MD, Portera-Cailliau C, Koliatsos VE (1995) Nerve growth factor prevents apoptotic cell death in injured central cholinergic neurons. J Comp Neurol 359: 573–585

Wisniewski K, Jervis GA, Moretz RC, Wisniewski HM (1979) Alzheimer neurofibrillary tangles in diseases other than senile and presenile dementia. Ann Neurol 5: 288–294

Yankner BA, Duffy LK, Kirschner DA (1990) Neurotrophic and neurotoxic effects of amyloid β protein: reversal by tachykinin neuropeptides. Science 250: 279–282

Yankner BA, Mesulam MM (1991) β amyloid and the pathogenesis of Alzheimer's disease. N Engl J Med 325: 1849–1857

Yoshiyama Y, Yamada T, Asanuma K, Asahi T (1994) Apoptosis related antigen, Le(Y) and nick-end labeling are positive in spinal motor neurons in amyotrophic lateral sclerosis. Acta Neuropathol 88: 207–211

Authors' address: Dr. C. Bancher, Ludwig Bolzmann Institute of Clinical Neurobiology, Department of Neurology, Lainz Hospital, Wolkersbergenstrasse 1, A-1130 Wien, Austria.

Assessment of neurotoxicity and "neuroprotection"

Muriel O'Byrne[1], K. Tipton[1], G. McBean[1], and H. Kollegger[2]

[1]Department of Biochemistry, Trinity College, Dublin, Ireland
[2]Neurological Clinic, University of Vienna, Vienna, Austria

Summary.   Coronal brain slices allow the study of neurotoxicity and "neuroprotection" under conditions where the differentiation-state and interrelationships of the neurones and glial cells are closer to those occurring in the intact tissue than is the case for co-cultured cell systems. The involvement of glial cells in the excitotoxicity of kainate and the potentiation of this toxicity by inhibition of glutamine synthase can be demonstrated. Longer-term toxicity of kainate may also be compounded by depletion of glutathione levels resulting from inhibition of γ-glutamylcysteine synthase. The involvement of nitric oxide formation in the toxicity of N-methyl-D-aspartate can also be shown. The neurotoxicity of 1-methyl-4-phenylpyridinium can be readily demonstrated in coronal slice preparations. Taurine affords protection against this neurotoxicity. The possible mechanisms of these effects are considered in terms of the cyclic interrelationships between the different events which can lead to cell death.

Introduction

There are several in vivo and in vitro approaches to the study of neurotoxicity and neuroprotection each of which has its own advantages and disadvantages. Coronal slices from the rat brain have several distinct advantages over the use of isolated cultured cells as a system in which to study the onset of neurodegenerative changes in brain tissue. The advantages of the tissue-slice model may be summarised as follows:

- It is possible to study neurodegenerative processes over a much shorter time-scale than is normally realisable with in vivo experiments. Previous studies have shown morphological evidence of degeneration of striatal neurones after as little as 40 minutes incubation with neurotoxic analogues of glutamate (see McBean, 1990).
- The involvement of neuronal-glial cell interactions in neurotoxic and neuro-protective processes can be studied in a situation where the interrelationships and differentiation state are similar to those occurring in the intact tissue. The use of cultured neurones or even peripheral models, such as PC-12 cells, is less satisfactory since the important interactions between

neuronal and glial cells are absent. This problem may be overcome, at least in theory, by the use of neurone-glial cell co-cultures. However, these suffer from the problems of uncertainties about their differentiation state which is, of course, a major problem when immortalised cell lines are studied. Evidence from enzyme localisation and receptor-interaction studies indicates that astrocytes are differentiated in such a manner that they complement and sustain the activity of specific neuronal systems.

- It is possible to prepare slices from the brains of animals of different ages so that the influence of age on the susceptibility to neurotoxicity may be studied. This contrasts with primary cultured neurones where it has not proven possible to prepare viable cultures from adult rats.
- It is possible to pretreat, by stereotaxic injection, with a possible neuroprotective agent and then study the "ex vivo" responses of the tissue slices to neurotoxins as an alternative to treating the isolated supervised slice with both the neuroprotective and neurotoxic agents in vitro. As discussed below, these two approaches have, for example, been shown to give different results in terms of the protection of neurones afforded by α-aminoadipate from kainate and N-methyl-D-aspartate (NMDA) toxicity.
- Neurotoxicity may be assessed by microscopic examination of the fixed and stained sections or by the use of specific cell-type markers or neurotransmitter levels.

Despite these advantages, there are also disadvantages of the slice model that need to be born in mind. Even with slices of 0.4–0.5 mm thickness, some degree of anoxia might be expected to affect the inner cellular layers, although microscopic examination does not indicate that any extensive necrosis occurs. The viability of superfused slices is relatively short and precludes studies of slow neurodegeneration arising from, for example, chronic neurotoxin exposure. A difficult-to-quantify degree of trauma might be expected to be associated with the preparation of the slices and this may affect their responses to toxins. Their restricted time of viability means that extensive "recovery times" cannot be used.

The cell types present in a brain slice preparation are much more varied than might normally be used in cultured cell systems. The coronal slices used in the work discussed below comprise neurones and glial cells from the striatum, corpus callosum and part of the cerebral cortex. The possibility that complex cell-cell interactions may affect the responses could make the data obtained difficult to compare directly with those obtained with simpler systems.

Results and discussion

Excitotoxins

Some examples of the assessment of excitotoxicity and protection by DL-α-aminoadipate (α-AA), DL-aminophosphonovalerate (APV), L-methionine sulphoximine (MSO) and L-N^G-nitroarginine (NNA) are summarised in Table 1.

Table 1. Effects of pretreatments on excitotoxicity in rat brain coronal slices

Toxin/Compound	Kainate	NMDA	Quinolinate
α-AA in vivo	Protection	No protection	No protection
α-AA in vitro	No protection	Protection	Protection
APV	No protection	–	Protection
MSO	Potentiation	–	Potentiation
NNA	No protection	Protection	–

In the in vitro studies the slices were exposed to kainate (300 μM), NMDA (500 μM) or quinolinate (500 μM) for 40 min in the presence and absence of DL-α-aminoadipate (α-AA; 1–3 mM) or DL-aminophosphonovalerate (APV; 1 mM). In the experiments with methionine sulphoximine (MSO; 500 μM) and L-N^G-nitroarginine (NNA; 100 μM) the slices were preincubated for 20 min with these compound before the addition of kainate or NMDA, at the above concentrations, and incubation for a further 20 min. All incubations were performed at 30°C in oxygenated Krebs bicarbonate medium. Neuronal cell viability was assessed microscopically after fixation and thionin staining and also by the determination of the levels of γγ-enolase (neurone-specific enolase, NSE) by radio-immunoassay. None of the protective agents used alone significantly affected neuronal survival. Results summarised from McBean (1990) and Kollegger et al. (1991 and 1993)

These results would be consistent with the following mechanisms:

1. The maintenance of normal glial-cell function is a contributory factor in kainate toxicity but not that of the other excitotoxins, NMDA and quinolinate. Preteatment with α-AA in vivo has been shown to impair glial cell function and morphology in a variety of systems (Lunkarlsen, 1979; Olney et al., 1971, 1982; Pedersen and Karlsen, 1979; McBean, 1990). Such effects are of temporary duration; direct intra-striatal injection of 100 μg α-AA (in a total volume of 2 μl), for example, caused a significant reduction of the activities of the glial cell enzyme glutamine synthase and of citrate synthase at 6 h after treatment but these values had returned to normal after 24 h. This treatment had no effects on the morphology of the striatal neurones. A similar selective and transient impairment of glial cell function has been reported after injection of the metabolic toxin fluorocitrate (Paulsen et al., 1987).
2. Pretreatment with α-AA in vitro protects against NMDA and quinolinate toxicity by virtue of its ability to act as an antagonist of the receptors for these two excitotoxins. There was no significant protection against kainate toxicity, thus excluding the possibility that the results obtained after intrastriatal injection of α-AA arise from direct receptor antagonism.
3. The glutamate receptor antagonist APV also protected against the neurotoxic responses to NMDA but not those of kainate.
4. The formation of nitric oxide, presumed to result from the activation of nitric oxide synthase as a consequence of elevated concentrations of intercellular free calcium ions, plays a role in the neurotoxicity of NMDA but not that of kainate.

5. The glutamine synthase inhibitor methionine sulphoximine potentiates the
   neurotoxic effects of both kainate and NMDA. This enzyme, which is
   exclusively localised in glial cells (Norenberg and Martinez-Hernandez,
   1979), has been proposed as a component of a "glutamine cycle" in which
   glutamate, released from neurones is captured by glial cells and converted
   to glutamine which is exported to the neurones and returned to the
   transmitter pool by the action of the neuronally-located phosphate-activat-
   ed glutaminase (Waniewski and Martin, 1986). There is, however, a difficulty with this theory, since glutamine is not efficiently taken up by
   neurones (Hertz et al., 1980). An alternative possibility is that the primary
   purpose of glutamine synthase in the glial cells may be for the removal of
   potentially toxic ammonia. Consistently with this it is the physiological
   concentration of ammonia, rather than that of glutamate, which regulates
   the activity of the enzyme (Waniewski, 1992). Furthermore, the potent
   convulsant properties of methionine sulphoximine have been ascribed to

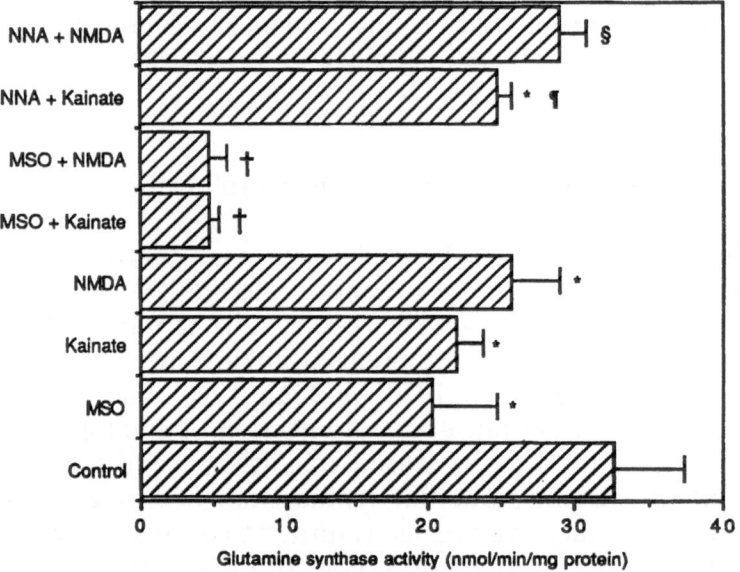

Fig. 1. Effects of kainate, N-methyl-D-aspartate (NMDA), L-methionine sulphoximine
(MSO) and L-N^G-nitroarginine (NNA) on the activity of glutamine synthase in rat brain
coronal slices. The slices were incubated at 30°C in oxygenated Krebs bicarbonate
medium for 60 min before the addition of kainate (300 μM), NMDA (500 μM) or MSO
(500 μM) and incubation for a further 40 min. In the experiments with MSO and
excitotoxin the slices were incubated with 500 μM MSO for 20 min before the addition of
500 μM kainate or 500 μM NMDA and a further 20 min incubation. In the experiments
with NNA and excitotoxin the slices were incubated with 100 μM NNA for 20 min before
the addition of 500 μM kainate or 500 μM NMDA and a further 40 min incubation.
Glutamine synthase activity was determined as the γ-glutamyltransferase activity (Wellner and Meister, 1966; Schousboe, 1982). Values are means ± SD (n = 5–7, each determined in duplicate). Statistical analysis was performed by the Neuman-Keuls test:
*p < 0.05 with respect to control (incubated with medium alone), †p < 0.05 with respect to
the excitotoxin alone, ¶NS with respect to kainate alone, §p < 0.05 with respect to NMDA
alone

the accumulation of ammonia, rather than a build-up of glutamate itself, and brain glutamate levels actually fall following MSO administration (Cooper et al., 1983). Although methionine sulphoximine inhibits several other brain enzymes, including γ-glutamylcysteine synthase, other studies with specific inhibitors and derivatives, have indicated that the convulsions induced by this compound specifically result from the inhibition of glutamine synthase activity (Cooper et al., 1983).

Incubation of the brain slices with either kainate or NMDA alone resulted in a decrease of glutamine synthase activity (Fig. 1) suggesting a direct or indirect effect of both these toxins on this aspect of glial cell function. The effects of kainate are consistent with the observation of Krespan et al. (1982) that this excitotoxin reduces glutamine concentrations in rat cerebellar slices, which these workers presumed to result from inhibition of glutamine synthase activity. Direct intra-hippocampal application of kainate has also been shown to reduce glutamine synthase activity, determined 24 hours later (Waniewski and McFarland, 1990).

However, direct studies with brain striatal homogenates indicated that kainate was not an inhibitor of this enzyme, even at concentrations as high as 5 mM (Fig. 2). Furthermore, the presence of 5 mM kainate in brain homogenates did not enhance the 37% inhibition of glutamine synthase activity given by methionine sulphoximine alone in such homogenates. This contrasts with the supra-additive decrease in glutamine synthase activity seen with pre-incubation of the slices with methionine sulphoximine prior to the addition of kainate (Fig. 1).

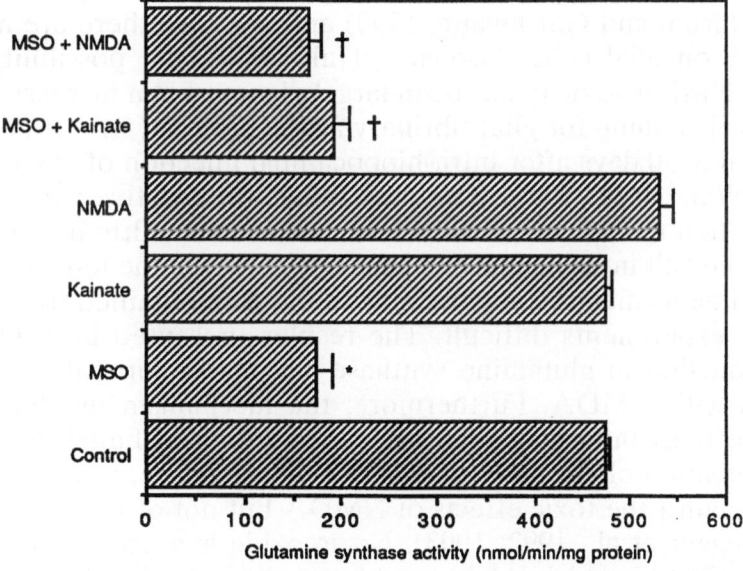

Fig. 2. Effects of kainate, N-methyl-D-aspartate (NMDA) and L-methionine sulphoximine (MSO) on the activity of glutamine synthase in rat brain striatal homogenates. Glutamine synthase activity was determined in 10% homogenates of striatal tissue in 0.1 M sodium acetate, pH 7.4. *p < 0.05 with respect to control (incubated with medium alone), †NS with respect to MSO alone

Kainate has been shown to cause a reduction in ATP levels, and a rise in extra-cellular glutamate concentration in the medium surrounding cerebellar slices (Krespan et al., 1982; Nicklas et al., 1980), which could arise by leakage from the cytoplasm by means of the carrier protein (Pocock et al., 1988). This could account for the release being calcium-independent (Ferkany et al., 1982). Since most of the transport sites for glutamate are present on glial cells (Drejer et al., 1982), it is possible that the initial effects of this toxin, like methionine sulphoximine may take place in glial cells (see McBean, 1990). The possible involvement of glial cell function in kainate toxicity might thus result from that excitotoxin causing release of glutamate from the glial cell pool. If that were the case, the potentiation afforded by methionine sulphoximine might result from that compound also functioning to release glial cell glutamate. However, it is extremely difficult to extrapolate data from the results of experiments such as these to the order of events that may occur in vivo, since they do not allow an easy distinction between the primary or secondary effects of a particular toxin. This aspect will be considered later in terms of the interrelationships between the different processes involved in cell death and neuroprotection.

As was the case with kainate, NMDA, at concentrations of up to 5 mM had no direct inhibitory effects on glutamine synthase activity in striatal homogenates and neither did it enhance the inhibition given by methionine sulphoximine alone. However, in contrast to the behaviour of kainate preincubation of the brain slices with NNA protected against the inhibition of glutamine synthase activity by NMDA (Fig. 1). These results suggest that the effects of NMDA on glutamine synthase activity are mediated by nitric oxide.

The mechanism of NMDA-induced neurodegeneration has generally been explained in terms of hyper-activation of post-synaptic receptors for glutamate (see Meldrum and Garthwaite, 1990) and, because there are no receptors for NMDA on glial cells (Usowicz et al., 1989), the possibility of a glial element in NMDA toxicity has been largely ignored. An increase in immuno-cytochemical staining for glial fibrillary acidic protein (GFAP), a marker for glial cells, at 3–30 days after intra-hippocampal injection of NMDA has been reported (Wang et al., 1991). This may reflect the reactive gliosis around the site of injection that occurs with such in vivo lesions. Although, this might be expected to result in an increase in glial cell enzymes, the longer time-scale of such experiments makes comparison with the data obtained from the present tissue-slice experiments difficult. The results, presented in Table 1, clearly show a reduction in glutamine synthase activity in coronal slices following incubation with NMDA. Furthermore, the mechanism involved is distinct from the decrease induced by kainate, since inhibition of nitric oxide synthase, by pre-incubation of the slices with L-N^G-nitroarginine, caused protection of the slices against the toxic effects of NMDA but not of kainate (Izumi et al., 1992; Kollegger et al., 1992, 1993). Nitric oxide is a product of nitric oxide synthase activation which is triggered by receptor-mediated entry of calcium into the post-synaptic cell. It has been described as a retrograde messenger, since it can readily diffuse from the site of synthesis to affect adjacent cells (see e.g. Garthwaite, 1991), where its primary target would appear to be activation of soluble guanylate cyclase (Garthwaite, 1991). Interestingly, Kiedrowski et al. (1992) have shown that cerebellar astrocytes, which do not express nitric oxide synthase, are nevertheless activated to produce cGMP by the nitric oxide that is produced in adjacent granule cells. This signalling mechanism would make it possible that, even indirectly, NMDA could influence glial cell metabolism in by a different mechanism from that involving kainate. The above results indicate that excitotoxic events may be more complex than is often envisaged; involving a variety of different factors and with the participation of glial cells in addition to the neurones themselves. The implications of the involvement of the glial cells in the processes of the neurotoxicity of both these compounds merit further study.

A further factor that may be relevant to the toxicity of kainate is that this compound has been shown to be an inhibitor of the enzyme γ-glutamylcysteine synthase in rat brain (McBean et al., 1995). The activity of this enzyme has been reported to be the major rate-limiting factor in the γ-glutamyl cycle, which is responsible for the maintenance of glutathione (GSH) levels within cells. These results suggest that kainate may cause a depletion of GSH and hence an impairment of free-radical detoxification mechanisms in brain tissue. Depletion of GSH has been shown to render the cell considerably more susceptible to the damage caused by free-radicals (Pellmar et al., 1992) and anti-oxidants can reduce some types of toxin-induced brain damage (see Meldrum and Garthwaite, 1990; Tipton, 1994). Activation of at least one subtype of glutamate receptor increases production of the superoxide anion in cell cultures (Lafon-Cazal et al., 1993) and since this may play a role in the neurodegenerative process, the possible contribution of glutathione depletion to the neuronal cell death induced by kainate requires further, more detailed, study.

Mitochondrial toxins

The neurotoxicity of 1-methyl-4-phenyl-1,2,3,6-tetrahydropyridine (MPTP) also involves glial cell function since it is the monoamine oxidase-B in those cells that catalysed the conversion of this compound to the effective toxin 1-methyl-4-phenylpyridinium (MPP$^+$). MPP$^+$ is an inhibitor mitochondrial ATP formation which is actively accumulated by dopaminergic neurones and it is believed that this inhibition of energy metabolism is sufficient to result in neurodegeneration (for review see Tipton and Singer, 1992). This is supported by the observation that direct intra-cerebral injection of the mitochondrial inhibitor rotenone, which like MPP$^+$ inhibits the NADH-ubiquinone reductase (Complex I) of the respiratory chain, also causes neurodegeneration (Heikkila et al., 1985).

As with other neurotoxins the behaviour of MPTP appears to be more complex than is suggested by this simple model. MPP$^+$ has been shown to induce Ca^{2+} release from mitochondria and an increase in Ca^{2+} cycling, resulting in collapse of membrane potential (Frei and Reichter, 1986). Disturbance of calcium homeostasis and increased levels of cytosolic calcium may therefore contribute to the nigrostriatal neuronal death. The inhibition of mitochondrial electron transport will, itself, result in the generation of oxygen radicals (see Hasegawa et al., 1990; Turrens and Boveris, 1980) which might also contribute to the neurodegenerative process. Indeed, antioxidants, such as glutathione and ascorbate, have been reported to afford and some protection against MPTP toxicity (Wagner et al., 1985; Cleeter et al., 1992; but see Tipton, 1994). In turn, oxidative stress itself would be expected to result in increased calcium ion efflux from mitochondria (Sandri et al., 1990).

Similar the neurotoxic action of 6-hydroxydopamine, which has been generally regarded as resulting from the generation of oxygen radicals (see e.g. Cohen and Werner, 1994), may also involve several different factors, since this compound is also a directly-acting inhibitor of mitochondrial respiratory chain Complex I (Glinka et al., 1996).

The neurotoxicity of MPP$^+$ can be readily demonstrated in the coronal slice preparation by the consequent reduction of the levels of neurotransmitter amines (O'Byrne et al., 1996) and of $\gamma\gamma$-enolase. In this system the sulpho-amino acid taurine (2-aminoethanesulphonic acid) has been shown to afford protection against the neurotoxicity of MPP$^+$. Taurine has been variously proposed to be an osmo-regulator, an antioxidant, a calcium-ion modulator, a membrane-stabiliser, a modulator of trophic factors and a neurotransmitter or neuromodulator (see e.g. Bianchi et al., 1994; Huxtable, 1992; Lombardini, 1991; Park et al., 1995; Schurr and Rigor, 1987). Clearly further work will be necessary to establish which of these factors underlies the neuroprotective effects of taurine in this system.

Conclusions

Although it is tempting to regard the process of neurotoxicity as being initiated by a single discrete insult that then triggers the chain of events that results in apoptotic cell death, the results presented here indicate that the process may be more complex, with an interrelated series of processes being involved in the final destruction of the cell. Thus, as referred to above, mitochondrial inhibition will result in elevation of calcium ions and formation of reactive oxygen radicals which, in turn, may cause membrane damage. Similarly the excitotoxin-induced elevation of cellular calcium ions will result in the formation of oxygen radicals and consequent mitochondrial and membrane damage. The fall in cellular ATP levels and release of glutamate which occur in anoxia will, of course, have similar interrelated consequences. Such interrelationships are illustrated in Fig. 3 which shows that a toxic insult affecting one component of the "cell-death cycle" will have knock-on effects on each of the other components, each of which can then contribute to the final chain of events leading to apoptosis (see e.g. Sen and D'Incalci, 1992). The above scheme is grossly oversimplified, since several other factors may also come into play. For example, localised iron-overload may result in those areas subject to toxic cell damage, perhaps resulting from activation of microglia, and further exacerbating oxygen radical formation (see Youdim and Riederer, 1993). Furthermore, as mentioned above, the inhibition of glutathione formation by toxins such as kainate and NMDA may render the cell more vulnerable to apoptotic death.

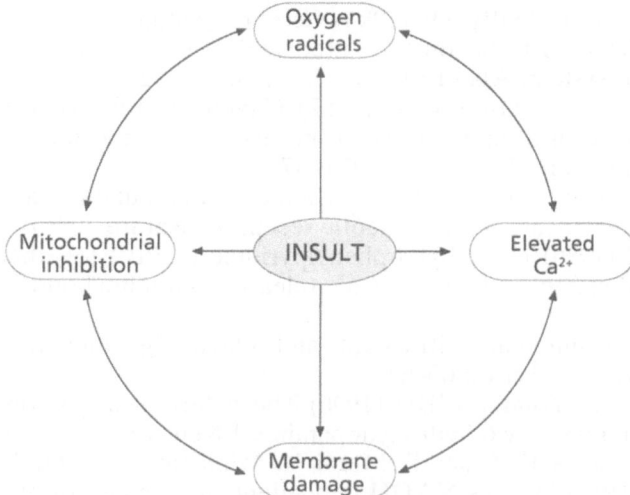

Fig. 3. The cell-death cycle in oversimplified form. Further discussion is given in the text

Leaving aside the importance of neuronal-glial cell interactions in the actions of at least some neurotoxins, that has been discussed above, and the possible induction of trophic factors, these considerations may have important consequences for interpreting the actions of the disparate range of compounds that have been reported to be neuroprotective. It may be that a single toxic event can predominate at high doses of an acutely-applied toxin, but in such cases it is possible that cell death may be predominately necrotic rather than apoptotic. At lower toxin levels these interrelationships may have important consequences for interpreting the actions of compounds that are neuroprotective. In terms of the "cell-death cycle" discussed above it follows that, just as this process of interrelated events may be initiated by a toxic event primarily affecting one component of the system, at least some measure of protection may result from the actions of compounds that affect the cycle at steps different from that which is the primary target of the toxic event.

Acknowledgements

We are grateful to the European Union (BIOMED 1 programme) and the Health Research Board for support.

References

Bianchi L, Sharp T, Bolam JP, Della Corte, L (1994) The effect of kainic acid on the release of GABA in rat neostriatum and substantia nigra. Neuro Report 5: 1233–1236

Cleeter MWJ, Cooper JM, Schapira AHV (1992) Irreversible inhibition of mitochondrial complex I by 1-Methyl-4-phenylpyridinium (MPP+); evidence for free radical involvement. J Neurochem 58: 786–789

Cohen G, Werner P (1994) Free radicals, oxidative stress and neurodegeneration. In: Calne DB (ed) Neurodegenerative disease. Saunders, Philadelphia, pp 139–162

Cooper AJL, Vergara F, Duffy TE (1983) Cerebral glutamine synthetase. In: Hertz L, Kvamme E, McGeer EG, Schousboe A (eds) Glutamate, glutamine and GABA in the central nervous system. Alan Liss, New York, pp 77–93

Drejer J, Larsson OM, Schousboe A (1982) Characterisation of L-glutamate and L-glutamine uptake into, and release from, astrocytes and neurones cultured from different brain regions. Exper Brain Res 47: 259–269

Ferkany JW, Zaczek R, Coyle JT (1982) Kainic acid stimulates excitatory amino acid neurotransmitter release at pre-synaptic receptors. Nature 298: 757–759

Frei B, Richter C (1986) N-methyl-4-phenylpyridine (MPP$^+$) together with 6-hydroxy-dopamine or dopamine stimulates Ca^{2+} release from mitochondria. FEBS Lett 198: 99–102

Garthwaite J (1991) Glutamate, nitric oxide and cell-cell signalling in the central nervous system. Trends Neurosci 14: 60–67

Glinka Y, Tipton KF, Youdim MBH (1996) The nature of inhibition of mitochondrial respiratory complex I by 6-hydroxydopamine. J Neurochem 66: 2004–2010

Hasegawa E, Takeshiga K, Oishi T, Murai Y, Minikami S (1990) 1-Methyl-4-phenyl-pyridinium (MPP$^+$) induces NADH dependent superoxide formation and enhances NADH-dependant lipid peroxidation in bovine heart submitochondrial particles. Biochem Biophys Res Commun 170: 1049–1055

Hertz L, Yu A, Svenneby G, Kvamme E, Fosmark H, Schousboe A (1980) Absence of a preferential glutamine uptake in neurons – an indication of a net transfer of TCA constituents from nerve endings to astrocytes? Neurosci Lett 16: 103–109

Heikkila RE, Nicklas WJ, Vyas I, Duvoisin RC (1985) Dopaminergic toxicity of rotenone and the 1-methyl-4-phenylpyridinium ion after their stereotaxic administration to rats: implications for the mechanism of 1-methyl-4-phenyl-1,2,3,6-tetrahydropyridine toxicity. Neurosci Lett 62: 389–394

Huxtable RJ (1992) Physiological actions of taurine. Phys Rev 72: 101–159

Izumi Y, Benz AM, Clifford DB, Zorumski CF (1992) Nitric oxide inhibitors attenuate N-methyl-D-aspartate excitotoxicity in rat hippocampal slices. Neurosci Lett 135: 227–230

Kiedrowski L, Costa E, Wroblewski JT (1992) In vitro interaction between cerebellar astrocytes and granule cells: a putative role for nitric oxide. Neurosci Lett 135: 59–61

Kollegger H, McBean GJ, Tipton KF (1991) The inhibition of glutamine synthetase in rat corpus striatum in vitro by methionine sulfoximine increases the neurotoxic effects of kainate and N-methy-D-aspartate. Neurosci Lett 130: 95–98

Kollegger H, McBean GJ, Tipton KF (1992) Striatal NMDA-neurotoxicity is reduced by inhibition of nitric oxide synthase. Neurochem Int 21 [Suppl]: B9

Kollegger H, McBean GJ, Tipton KF (1993) Reduction of striatal N-methyl-D-aspartate toxicity by inhibition of nitric oxide synthase. Biochem Pharmacol 45: 260–264

Krespan B, Berl S, Nicklas WJ (1982) Alteration in neuronal-glial metabolism of glutamate by the neurotoxin kainic acid. J Neurochem 38: 509–518

Lafon-Cazal M, Pietri S, Culcasi M, Bockaert J (1993) NMDA-dependent superoxide production and neurotoxicity. Nature 364: 535–537

Lombardini JB (1991) Taurine and retinal function. Brain Res Rev 16: 151–169

Lunkarlsen R (1979) The toxic effects of sodium glutamate and DL-α aminoadipic acid on rat retina: changes in high affinity uptake of putative neurotransmitters. J Neurochem 31: 1055–1061

McBean GJ (1990) Intrastriatal injection of DL-α aminoadipate reduces kainate toxicity in vitro. Neurosci 34: 225–234

McBean GJ, Horner EB, Couée I, Phillips JP, O'Brien M, Lee TC, Tipton KF (1990) Enzymes and glial cells in brain damage and neurodegenerative diseases. In: Dostert P, Riederer P, Strolin Benedetti M, Roncucci R (eds) Early markers in Parkinson's and Alzheimer's disease (New vistas in drug research 1). Springer, Wien New York, pp 209–220

McBean GJ, Doorty KB, Tipton KF, Kollegger H (1995) Alterations in the glial cell metabolism of glutamate by kainate and N-methyl-D-aspartate. Toxicon 33: 569–576

Meldrum B, Garthwaite J (1990) Excitatory amino acid neurotoxicity and neurodegenerative disease. Trends Pharm Sci 11: 379–387

Nicklas WJ, Krespan B, Berl S (1980) Effect of kainate on ATP levels and glutamate metabolism in cerebellar slices. Eur J Pharmacol 62: 209–213

Norenberg MD, Martinez-Hernandez A (1979) Fine structural localisation of glutamine synthetase in astrocytes of rat brain. Brain Res 161: 301–310

O'Byrne MB, Tipton KF, McBean GJ (1996) Neurotoxicity of MPP$^+$ as studied in a tissue slice system. Biochem Soc Trans 24: 62S

Olney JW (1982) The toxicity of glutamate and related compounds in the retina and the brain. Retina 2: 341–359

Olney JW, Ho OL, Rhee V (1971) Cytotoxic effects of acidic and sulphur-containing amino acids on the infant mouse central nervous system. Exper Brain Res 14: 61–76

Park E, Schuller-Levis G, Quinn MR (1995) Taurine chloramine inhibits production of nitric oxide and TNF-α in activated RAW 264.7 cells by mechanisms that involve transcriptional and translational events. J Immunol 154: 4778–4784

Paulsen RA, Contestabile A, Villani A, Fonnum F (1987) An in vivo model for studying function of brain tissue temporarily devoid of glial cell metabolism: the use of fluorocitrate. J Neurochem 48: 1377–1385

Pedersen OO, Karlsen RL (1979) Destruction of Muller cells in the adult rat by intravitreal injection of DL-α aminoadipic acid An electron microscopic study. Exper Eye Res 28: 569–575

Pellmar TC, Roney D, Lepinski, DL (1992) Role of glutathione in repair of free radical damage in hippocampus in vitro. Brain Res 583: 194–200

Pocock JM, Murphie HM, Nicholls DG (1988) Kainic acid inhibits the synaptosomal plasma membrane glutamate carrier and allows glutamate leakage from the cytoplasm but does not affect glutamate exocytosis. J Neurochem 50: 745–751

Sandri G, Panfili, E, Ernster, L (1990) Hydrogen peroxide production by monoamine oxidase in isolated rat brain mitochondria: its effect on glutathione levels and Ca^{2+} efflux. Biochim Biophys Acta 1035: 300–305

Schousboe A (1982) Glial cell marker enzymes. Scand J Immunol 15: 339–356

Schurr A, Rigor BM (1987) The mechanism of neuronal resistance and adaptation to hypoxia. FEBS Lett 224: 4–8

Sen S, D'Incalci M (1992) Apoptosis: biochemical events and relation to cancer chemotherapy. FEBS Lett 307: 122–127

Tipton KF (1994) What is it that I-deprenyl selegiline might do? Clin Pharmacol Ther 56: 781–796

Tipton KF, Singer TP (1993) Advances in our understanding of the mechanisms of the neurotoxicity of MPTP and related compounds. J Neurochem 6: 1191–1206

Turrens JF, Boveris A (1980) Generation of superoxide anion by the NADH dehydrogenase of bovine heart mitochondria. Biochem J 191: 421–427

Usowicz MM, Gallo V, Cull-Candy SG (1989) Multiple conductance channels in type 2 cerebellar astrocytes activated by excitatory amino acids. Nature 339: 380–383

Wagner GC, Jarvis MF, Carelli RM (1985) Ascorbic acid reduced dopamine depletion induced by methamphetamine and 1-methyl-4-phenylpyridinium ion. Neuropharmacol 24: 1261–1262

Wang S, Lees GJ, Rosengren LE, Carlsson J-E, Stigbrand T, Hamberger A, Haglid KG (1991) The effect of N-methyl-D-aspartate lesion in the hippocampus on glial and neuronal marker proteins. Brain Res 541: 334–341

Waniewski RA (1992) Physiological levels of ammonia regulate glutamine synthesis from extracellular glutamate in astrocyte cultures. J Neurochem 58: 167–174

Waniewski RA, Martin DL (1986) Exogenous glutamate is metabolised to glutamine and exported by rat primary astrocyte cultures. J Neurochem 47: 304–313

Waniewski RA, McFarland D (1990) Intrahippocampal kainic acid reduces glutamine synthetase. Neurosci 34: 305–310

Wellner VP, Meister A (1966) Binding of adenosine triphosphate and adenosine diphosphate by glutamine synthetase. Biochemistry 5: 872–879

Youdim MBH, Riederer P (1993) The role of iron in senescence of dopaminergic neurons in Parkinson's disease. J Neural Transm [Suppl] 40: 57–67

Authors' address: Prof. Dr. K. Tipton, Department of Biochemistry, Trinity College, Dublin 2, Ireland.

Immunoinflammatory mechanisms, infective diseases causing neurological disorders

Chair: R. Horowski and **W. Poewe**

Update on management and genetics of multiple sclerosis

A. Dessa Sadovnick

Department of Medical Genetics, University of British Columbia, Vancouver, Canada

Summary. Although the exact etiology of multiple sclerosis (MS) remains uncertain, there is an increasing body of evidence to support the role of genetic factors in MS susceptibility in general and the familial aggregation of MS in particular.

MS management continues to be largely symptom-specific. MS relapses are now frequently treated with IV-methylprednisone. In recent years, MS treatment trials are being conducted throughout the world. Interferon beta-1b has been approved as the first ongoing therapy for relapsing/remitting MS, although some issues/concerns remain to be addressed.

In summary, while much research is still needed, important progress is being made in unravelling the etiology of MS and in developing management protocols which are not symptom- or relapse-specific.

MS overview

Literary descriptions of MS have existed for centuries. The diary of Sir Augustus Frederic d'Este (1794–1848) is probably the best known of these (Firth, 1948). His diary described a disorder that spanned 25 years of his life and was characterized by visual impairment, numb feet, gait and balance problems, weakness and bladder dysfunction. The first pathological reports appear to be those by Carswell (1838) and Cruveilhier (1835). It was, however, Charcot, who recognized MS as a specific and recognizable entity, separating it from other neurological disorders such as paralysis agitans and neurosyphilis (Dejong, 1970). Eichhorst (1913) first labelled MS as an "inherited, transmissible" disease. Thus, although MS is not a "new" disease, we still cannot successfully treat or prevent this debilitating disorder. The cause of MS remains unclear although the role of genetic factors in MS susceptibility in general (Ebers and Sadovnick, 1994) and the familial aggregation of MS in particular (Ebers et al., 1995) appears to be important.

The age of onset for MS can range from 10 to 59 years (Poser et al., 1983) although MS has been reported to sometimes onset in early childhood (Duquette et al., 1992) or after age 59 (Noseworthy et al., 1983). Females are affected approximately twice as often as males (Weinshenker et al., 1989). Although MS has been reported worldwide, it is largely considered to affect caucasians of northern and central European ancestry (see Sadovnick and Ebers, 1994, for a review). While the clinical course of MS varies, at least 20% of affected individuals have "benign" disease (Marsden and Fowler, 1989), which is characterized by few or no exacerbations and little or no residual disability after a 10-year period. Natural history data suggest that 15 years after the initial onset of MS, 50% of patients can walk unaided, with the other 50% requiring a variety of walking aids (Weinshenker et al., 1989).

MS management

MS can be characterized by less obvious symptoms such as fatigue and mood disorders which, although not necessarily apparent to the casual observer, can greatly alter the affected individual's ability to conduct activities of daily living such as work, housekeeping, childcare, etc. Individual symptom management (e.g. spasticity, bladder dysfunction, fatigue, depression, infections, etc, the success of which vary, has long been the realistic goal of "treating" MS. MS relapses are now often treated with IV methylprednisone (Schapiro, 1994). Recently, the Food and Drug Administration (United States) and the Health Protection Branch (Canada) have approved interferon beta-1b (Betaseron®) as the first recognized therapy for relapsing-remitting MS (IFNB Multiple Sclerosis Study Group, 1995), although some issues/concerns must still be researched and resolved – see Table 1. Nevertheless, the results of the B-interferon trials have provided encouragement to researchers, MS patients and their families.

Genetics of MS

Evidence against a purely transmissible etiology in MS is strong, based on several types of observations including the low concordance rate in dizygotic twins (Ebers et al., 1986; Sadovnick et al., 1993), low rate of MS among spouses (Schapira et al., 1963; Finelli, 1991; Kaufman, 1992), the relatively

Table 1. Some issues to be researched and resolved with respect to B-interferon treatment in MS

- Initial side effects include flu-like illness and increased liver enzymes
- Repeated injections required
- MRI changes do not necessarily correlate with clinical signs and symptoms; major changes with B-interferon have been shown on MRI
- The impact of antibody formation must be further assessed and understood
- Safety to a fetus is unknown
- Longterm effect, e.g. to the immune system, in a child/adult whose mother took B-interferon during pregnancy is unknown
- Optimal "start time" and "end time" of treatment must be researched and further understood

---

high rate of MS in first-, second- and third-degree relatives of patients compared with the general population (Sadovnick et al., 1988; Ebers et al., 1995), the identification of groups resistant to MS in "high risk areas", e.g. Amerindians in North America (Hader et al., 1985), Japanese in the United States (Detels et al., 1977), Lapps in Scandinavia (Gronning and Mellgren, 1985 and the Yakutes in the far north of Russia (Popov, 1983), studies of age of onset versus year of onset in sibs concordant for MS (Bulman et al., 1991), and birth order position studies (Gaudet et al., 1995).

As highlighted by the "time line" given in Table 2, there has been increasing evidence that genetic factors have a role in determining susceptibility to MS. Re-examination of results from prevalence and migration surveys reveals that there remains considerable ambiguity in interpretation (see Sadovnick and Ebers, 1993, for a review). Some patterns previously thought to decisively support environmental determination may still be explained, at least in part, on a genetic basis. Nevertheless, data from the Antipodes (Hammond et al., 1988) still clearly show latitudinal differences in MS prevalence, despite relative homogeneity of the caucasian population. However, in general, it seems inescapable that MS is probably due to an interaction of genetic and non-genetic (environmental) factors.

It remains undetermined whether or not genes exist which are truly necessary for the development of the disease. Existing data suggest that the study of MS susceptibility will parallel the findings in experimental models of spontaneous autoimmunity and that, at the very least, two genes (and almost certainly several genes) will be found to influence MS susceptibility and interact in as yet unknown ways (Ebers and Sadovnick, 1994). One of these loci appears to be Class II MHC, although its role appears to be minor at the germ line level. Roles for the T-cell receptor alpha and beta loci are dubious. Existing data strongly indicate that additional loci will be identified which influence both susceptibility and outcome.

As part of the Canadian Collaborative Project on Genetic Susceptibility to MS (CCPGSMS), a population-based sample of 16,000 individuals with MS were screened (Sadovnick et al., 1996). Standardized, personally administered

Table 2. "Time Line": Genetics of multiple sclerosis

1868	–	First medical description of MS by Charcot (Dejong, 1970)
1896	–	Eichhorst labelled MS an "inherited, transmissible" disease (Eichhorst, 1913)
1967	–	Schumacher Diagnostic Criteria for MS (Schumacher et al., 1965)
1972	–	HLA association with MS first reported (Jersild et al., 1972)
1983	–	Poser Committee Diagnostic Criteria (Poser et al., 1983)
1986	–	Canadian multi-centre population based twin study in MS (Ebers et al., 1986)
1988	–	First published age-corrected familial risks for MS (Sadovnick et al., 1988)
1989	–	First International Workshop on the Genetic Susceptibility to MS
1995	–	Adoptee/adopted studies clearly show that the familial aggregation of MS is genetic (Ebers et al., 1995)

questionnaires were used to identify adopted MS patients and/or those who had adopted/adopting relatives. There was one case of MS among 1201 first-degree non-biological relatives of the MS patients, significantly less than the expected 25.4 cases expected based on data for first-degree biological relatives – $P = 2.5 \times 10^{-10}$ (Ebers et al., 1995). These findings indicate that the familial aggregation of multiple sclerosis is genetically determined and that no effect of shared familial environment was detectable (Ebers et al., 1995).

Results

In summary, while the exact cause of MS remains unclear and effective treatment for the overall disease is not yet available, research is progressing rapidly. Data now indicate that the familial aggregation is genetic and the role of genetic factors in MS susceptibility is increasingly recognized. Nevertheless, familial risk data, e.g. monozygotic twin concordance (Sadovnick et al., 1995), and information on the geographic distribution of the disease according to latitude, e.g. Australia (Hammond et al., 1988) lead to the important conclusion that the non-genetic (environmental) factors act in MS at a population level, rather than within families or at an "individual" level. These results have important implications for the future direction of MS research. Such research will be done as part of the CCPGSMS-Phase 2.

Acknowledgements

This research was funded by the Multiple Sclerosis Society of Canada Scientific Research Foundation, the Multiple Sclerosis Society of Canada and the the Multiple Sclerosis Society of Canada, British Columbia Division.

Drs. A.D. Sadovnick and G.C. Ebers (Department of Clinical Neurological Sciences, University of Western Ontario) are prinicipal investigators of the Canadian Collaborative Project on Genetic Suceptibility to MS. Dr. N. J. Risch (Department of Genetics, Stanford University) is the co-investigator.

Additional members of the Canadian Collaborative Study Group are: *London, Ontario* – D. Bulman, G.P.A. Rice; *Vancouver, British Columbia* – S.A. Hashimoto, D. Paty, J. J.-F. Oger, M.D.; *Calgary, Alberta* – L. Metz, R. Bell, M.D.; *Edmonton, Alberta* – S. Warren, Ph.D.; *Saskatoon, Saskatchewan* – W. Hader; *Winnipeg, Manitoba* – T. Auty, A. Nath; *Toronto, Ontario* – T. Gray, P. O'Connor; *Ottawa, Ontario* – R. Nelson, M. Freedman; *Kingston, Ontario* – D. Brunet; *Hamilton, Ontario* – R. Paulseth; *Montreal, Quebec* – G. Francis, (Montreal Neurological Institute) and P. Duquette, (Hopital Notre Dame); *Halifax, Nova Scotia* – T. J. Murray, V. Bahn; *St. John's, Newfoundland* – W. Pryse-Phillips.

This research was funded by the Multiple Sclerosis Society of Canada Scientific Research Foundation and the Multiple Sclerosis Society of Canada.

The following MS Clinic/Site Study Coordinators were responsible for accurate data collection: C. Harris; N. Cheyne; M. Hader; B. Davis; A. Royal; M. Perera; M. Penman; K. Stevenson; L. Murray; G. Smith; C. Edgar; J. Haynes; D. Southwell; R. Arnaoutelis; C. Laforce; C. Masse; P. Weldon; K. Taylor; G. Alcock. We would like to specially thank Holly Armstrong and Janna Smith (London), Rochelle Farquhar, Laila Mashal, Irene Yee, Dave Dyment, Thomas Hicks, Laila Albuhusan, Leanne McIntyre and Nancy Greig (Vancouver) for their ongoing assistance.

References

Bulman DE, Sadovnick AD, Ebers GC (1991) Age of onset in siblings concordant for multiple sclerosis. Brain 114: 937–950

Carswell R (1838) Pathological anatomy: illustrations of the elementary forms of disease. Longman, London

Cruveilhier J (1835) Anatomie pathologique du corps humain. JB Balliere, Paris

Dejong RN (1970) Multiple sclerosis: history, definition and general considerations. In: Vinken GW, Bryun PJ (eds) Handbook of clinical neurology. Elsevier, New York, pp 49

Detels R, Visscher BR, Malmgren RM, Coulson AH, Lucia MV, Dudley JP (1977) Evidence for lower susceptibility to multiple sclerosis in Japanese Americans. Am J Epid 105: 303–310

Duquette P, Pleines J, Girard M, Charest L, Senecal-Quevillon M, Masse C (1992) The increased susceptibility of women to multiple sclerosis. Can J Neurol Sci 19: 466–471

Ebers GC, Bulman DE, Sadovnick AD, et al (1986) A population-based twin study in multiple sclerosis. N Engl J Med 315: 1638–1642

Ebers GC, Sadovnick AD (1994) Genetics of multiple sclerosis: a critical overview. J Neuroimmunology 54: 117–122

Ebers GC, Sadovnick AD, Risch NJ, the Canadian Collaborative Study Group (1995) Familial aggregation in multiple sclerosis is genetic. Nature 377: 150–151

Eichhorst H (1913) Multiple Sklerose und spastische Spinalparalyse. Med Klin 9: 1617–1619

Finelli PF (1991) Conjugal multiple sclerosis: a clinical and laboratory study. Neurol 41: 1320–1321

Firth D (1948) The case of Augustus d'Este. Cambridge University Press, London

Gaudet JPC, Hashimoto L, Sadovnick AD, Ebers GC (1995) A study of birth order and multiple sclerosis in multiplex families. Neuroepidemiology 14: 188–192

Gronning M, Mellgren SI (1985) Multiple sclerosis in the two northernmost counties of Norway. Acta Neurologica Scandinavica 72: 321–327

Hader WJ, Feasby TE, Noseworthy JH, Rice GPA, Ebers GC (1985) Multiple sclerosis in Canadian native people. Neurol 35 [Suppl 1]: 300

Hammond SR, McLeod JG, Millingen KS, et al (1988) The epidemiology of multiple sclerosis in three Australian cities: Perth, Newcastle and Hobart. Brain 111: 1–25

IFNB Multiple Sclerosis Study Group and the University of British Columbia MS/MRI Analysis Group (1995) Interferon beta-1b in the treatment of multiple sclerosis: final outcome of the randomized controlled trial. Neurol 45: 1277–1285

Jersild C, Svejgaard A, Fog T (1972) HLA antigens and multiple sclerosis. Lancet 1: 1240–1241

Kaufman MD (1992) Conjugal multiple sclerosis. Neurol 42: 1644–1645

Marsden CD, Fowler TJ (1989) Demyelinating diseases of the central nervous system. In: Clinical neurology. Raven Press, New York, Chapter 16

Noseworthy J, Paty DW, Wonnacott T, et al (1983) Multiple sclerosis after age 50. Neurol 33: 1537–1544

Popov VS (1983) Clinical picture and epidemiology of disseminated sclerosis. Zh Nevropatol Psikhiatr 83: 1330–1334

Poser CM, Paty DW, Scheinberg L, et al (1983) New diagnostic criteria for multiple sclerosis: Guidelines for research protocols. Ann Neurol 13: 227–231

Sadovnick AD, Baird PA, Ward RH (1988) Multiple sclerosis: updated risks for relatives. Am J Med Genet 29: 533–541

Sadovnick AD, Ebers GC (1993) Epidemiology of multiple sclerosis: a critical overview. Can J Neurol Sci 20: 17–29

Sadovnick AD, Armstrong H, Rice GPA, et al (1993) A population-based twin study of multiple sclerosis in twins: update. Ann Neurol 33: 281–285

Sadovnick AD, Ebers GC, the Canadian Collaborative Study Group (1996) Basic, clinical and genetic epidemiology of MS. In: Abramsky O, Ovadia A (eds) Frontiers in multiple sclerosis: clinical research and therapy. Martin Dunitz Publications, London (in press)

Schapira K, Poskanzer DC, Millar H (1963) Familial and conjugal multiple sclerosis. Brain 86: 315–332.

Schapiro RT (1994) Symptom management in multiple sclerosis. Demos Publications Inc, New York

Schumacher GA, Beebe G, Kibler RF, et al (1965) Problems of experimental trials of therapy in multiple sclerosis. Report by the panel on the evaluation of experimental trials of therapy in multiple sclerosis. Ann NY Acad Med 122: 552–568

Weinshenker BG, Bass B, Rice GP, et al (1989) The natural history of multiple sclerosis: a geographically-based study. I. Clinical course and disability. Brain 112: 133–146

Author's address: Dr. A. Dessa Sadovnick, Department of Medical Genetics, University of British Columbia, # 222 Wesbrook Building, 6174 University Boulevard, Vancouver, B.C. V6T 1Z6, Canada.

Pathogenesis of immune-mediated demyelination in the CNS

H.-P. Hartung and **P. Rieckmann**

Department of Neurology, Julius-Maximilians-Universität, Würzburg,
Federal Republic of Germany

Summary. Collective evidence from studies in the animal model experimental autoimmune encephalomyelitis and pathological and immunological studies on MS patients suggest that this most common inflammatory demyelinating disorder of the central nervous system results from primarily T-lymphocyte driven aberrant immune responses to a number of myelin and possibly non-myelin antigens. These include MBP, PLP, MOG, MAG, CNP and S 100. Autoreactive T-cells reactive with these antigens circulate in blood and upon activation can travel across the blood-brain-barrier to initiate a local immunoflammatory response provided they encounter a microglial cell that displays antigenic epitopes in the context of MHC class II gene products and accessory molecules. Demyelination probably results from antibody-induced complement activation. Repeated inflammatory episodes eventually exhaust the reparative capacities of oligodendrocytes and damage axons. As the disease evolves, an initialy focussed immune response may diversify due to a process termed epitope spreading. The initial event of T lymphocyte activation remains elusive, but molecular mimicry, cross-recognition of structures shared between microbes and myelin, appears to be crucial.

Introduction

Multiple sclerosis in the majority of patients follows a relapsing remitting course. After repeated attacks, there is frequently a transition to a chronic progressive accumulation of neurological deficit. Pathologically, acute exacerbations are characterized by multifocal perivascular infiltration by mononuclear cells and demyelination of axons. With progression of the disease oligodendrocytes die and astroglial cells proliferate to give rise to gliotic scar formation (Raine, 1994). The following short review attempts to outline recent advances in our knowledge of the underlying pathogenesis of this disorder.

The role of T lymphocytes

Collective evidence gathered in experimental models and studies of MS patients indicates that MS results from a primarily T cell driven aberrant immune response to a number of myelin and possibly non-myelin antigens of the central nervous system. T cell autoreactivity detectable in blood and CSF is focussed on MBP (myelin basic protein), PLP (proteolipid protein), MOG (myelin oligodendrocyte glycoprotein), MAG (myelin associated glycoprotein) and the calcium binding protein S100 (Hohlfeld et al., 1995; Allegretta et al., 1990; Steinman et al., 1995). The presence and persistence of autoreactive T lymphocytes suggests that normal mechanisms maintaining self-tolerance have failed. These include physical deletion in the thymus, peripheral anergy e.g. functional silencing of circulating T cells in the blood with ensuing programmed cell death and active suppression by a specialized subpopulation of T lymphocytes (Theofilopoulos, 1995). There is evidence from work on MS plaque material and the animal model EAE (experimental autoimmune encephalomyelitis) that much of the T cell autoreactivity encountered in MS patients could be accounted for by the escape from peripheral anergy. T cell activation usually requires an antigen-specific signal to be transduced via the T cell receptor CD3 CD8/CD4 complex interacting with MHC class I or II molecules and additional co-stimulatory signals. The co-stimulatory signals are delivered during the interaction of a group of accessory molecule ligand pairs, B7-1/CD28 and B7-2/CTLA4. If these co-stimulatory signals are lacking, T-cells will not be activated but succumb to apoptosis. Increased local availability of such co-stimulatory signals, conversely, prevents anergy to occur and allows for persistence of activated autoreactive T-cells. In brains of patients with MS B7-1 was detected on activated microglia, infiltrating macrophages and perivenular lymphocytes (Windhagen et al., 1995; De Simone et al., 1995). EAE could be prevented by blocking B7-1/CD28 interactions (Kuchroo et al., 1995; Racke et al., 1995; Cross et al., 1995).

However, autoreactive myelin-specific T-cells can also be readily recovered from healthy individuals. Disease specificity may be attributed to: distinct recognition of a small relatively select set of immunodominant epitopes in the topographical context of similarly restricted MHC class II contact residues; utilization of a restricted set of T-cell receptor alpha and beta chain genes; an increased precursor frequency and a heightened state of activation in vivo (Wucherpfennig and Hafler, 1995; Markovic Plese et al., 1995; Wucherpfennig et al., 1994, 1995; Steinman et al., 1995).

The majority of T cells carry the αβ T cell receptor. In MS lesions significant numbers of T cells expressing γδ T cell receptors can also be found (Birnbaum, 1995). Some groups reported γδ T cell receptor restriction with clonal expansion which may be a finger print of an immune reponse against heat shock proteins (Birnbaum, 1995; van Noort et al., 1995; Wucherpfennig et al., 1992; Selmaj et al., 1991). Other than that, they may also play a pathogenic role by damaging oligodendrocytes (Battistini et al., 1995; Stinissen et al., 1995; Wucherpfennig et al., 1992; Freedman et al., 1991).

As will be outlined below, T cells in most likelihood are not the effector cells of tissue damage in MS. Rather, by elaborating cytokines, they recruit resident and immigrated mononuclear phagocytes (macrophages/microglia) to attack the myelin sheath and its cellular source, the oligodendrocyte. While early in the course of the disease, T cell responses may be restricted in terms of epitope recognition, interaction with MHC class II contact sites, utilization of TCR-α β chain genes, evidence from animal models suggests that repeated insults on CNS structures leading to a release of previously cryptic antigens broadens the aberrant immune response (a process termed epitope or determinant spreading) and set up a vicious circle (Miller et al., 1995; Lehmann et al., 1993).

The role of B lymphocytes

It has long been known that blood and CSF of MS patients contain autoantibodies derived from oligoclonal B cells. Interestingly, some of them recognize the same protein antigens as T cells. Indeed, at least in the case of MBP B cell and T cell epitopes appear to overlap (Warren et al., 1995).

T cell homing and migration

Systemic autoreactivity does not necessarily cause disease. Aberrant immune reponses must be focussed into the target organ, e.g. the brain and spinal cord. Only activated T cells can penetrate the blood-brain barrier and this process of activation may involve molecular mimicry, the presence of so-called superantigens or triggering cytokines generated in the course of a viral infection. Molecular mimicry implies a misguided immune response secondary to crossrecognition of epitopes shared between a microbial pathogen and the putative autoantigen in the central nervous system. There is some evidence that crossreactive amino acid sequences are shared between neurotropic viruses such as measles, influenza, adenovirus, herpes vrisus, Epstein-Barr virus and MBP and that MBP-reactive T cell clones derived from MS patients can be stimulated by these peptides (Wucherpfennig, Strominger, 1995).

A complex sequential interaction of adhesion molecules reciprocally expressed on the surface of leucocytes and endothelial cells allows myelinspecific T lymphocytes to home to the CNS (Hartung et al., 1995a; Butcher and Picker, 1996a). A footmark of the migratory process are increased concentrations of soluble adhesion molecules such as ICAM-1, VCAM-1 and L-selectin in serum and CSF of MS patients. These are associated with impaired integrity of the blood-brain-barrier (Rieckmann et al., 1994b; Hartung et al., 1993b, 1995a, 1995b; Tsukada et al., 1993b; Sharief et al., 1993b). Matrix metalloproteinases are crucially important for T cells to pass through basement membranes. Type IV gelatinase was found to open the blood-brainbarrier and selective pharmacological blockade of the active site of gelatinase abrogated inflammation in EAE (Gijbels et al., 1994). Raised levels of gelatinases have been measured in CSF from MS patients (Rosenberg et al., 1996; Gijbels et al., 1992).

The local immunoinflammatory cascade

Once autoreactive T cells have passed the blood-brain barrier they need to be reactivated in situ requiring again the trimolecular interaction with the respective antigenic epitopes and MHC II gene products. In addition, co-stimulatory molecules, e.g. B7-1 and IL-12, should be present in the local environment (Windhagen et al., 1985). Under these conditions immigrated autoreactive T cells can undergo proliferation and through release of cytokines start a cascade of inflammatory reactions (Olsson, 1995; Hartung et al., 1995a). CD4+ T helper cells, based on the pattern of cytokines they secrete, can be divided into Th1 and Th2 populations. Th1 cells synthesize IFN-γ, IL-2 and TNF-α and β while Th2 cells produce IL-4, IL-5, IL-6, IL-10 and IL-13. The Th1 cytokine tumor necrosis factor α causes myelin vesiculation in organotypic cultures (Selmaj and Raine, 1988) and worsens EAE in Lewis rats (Kuroda and Shimamoto, 1991). Antibodies to TNF-α abrogated demyelination in MBP T cell line-mediated EAE of mice (Selmaj et al., 1991b; Ruddle et al., 1990b). Serum and CSF concentrations have been reported to be elevated in patients with actively progressing MS (Hauser et al., 1990; Trotter et al., 1991; Sharief and Hentges, 1991) and more recently, a significant rise in TNF-α production by blood mononuclear cells has been documented to immediately antedate acute clinical exacerbations (Rieckmann et al., 1994b; Chofflon et al., 1992b; Beck et al., 1988b). Antigen-activated MBP-specific T cell clones from HLA-DR2+ donors secrete significantly more lymphotoxin than cells from HLA-DR2 negative patients (Zipp et al., 1995) emphasizing that cytokine production in MS is under genetic control.

TNF-α mediates its biologic effects through binding to distinct specific cell surface receptor types of 55 and 75 kDa molecular size (Tracey and Cerami, 1993). Increased concentrations of the 60 kDa TNF receptor can be measured in the serum of patients with MS (Hartung et al., 1995b; Rieckmann et al., 1994b; Matsuda et al., 1994b). In MS brains, TNF-α could be immunocytochemically localized to active plaques (Selmaj et al., 1991c; Hofman et al., 1989c). Both TNF receptors are expressed on oligodendrocytes and can be detected in active lesions (Sippy et al., 1995; Tchélingérian et al., 1995).

The second important Th1 inflammatory cytokine is interferon-γ. In a longitudinal study of MS patients a significant relation was observed between clinical attacks and changes in mitogen-driven synthesis of IFN-γ and TNF-α by peripheral blood leukocytes (Beck et al., 1988). There is also evidence of increased IFN-γ production by circulating T cells from MS patients (Link et al., 1994; Voskuhl et al., 1993; Olsson et al., 1990). Importantly, in a clinical trial systemic administration of IFN-γ to MS patients resulted in a higher frequency of clinical exacerbations, increased numbers of HLA-DR expressing circulating monocytes, enhanced proliferative responses of peripheral blood T cells and augmented NK cell activity (Panitch et al., 1987). IFN-γ has also been reported to activate an influx of calcium across the plasmalemma of T cells from relapsing-remitting MS patients, triggering T cell proliferation (Martino et al., 1995).

Inflammatory effector molecules

Macrophages/microglia can actively phagocytose myelin or injure the myelin sheath through a number of highly potent noxiuos molecules (Hartung et al.,

1992). Both reactive oxygen species and nitric oxide metabolites have been implicated in this regard. Oxygen radicals can cause lipid peroxidation and degradation of myelin in vitro (Konat and Offner, 1983; Konat and Wiggins, 1985; Chia et al., 1983) and fuel the nitric oxide pathway by generating peroxynitrite. Messanger RNA for nitric oxide synthase-2 (NOS-2) was detectable in MS brains localized to the cytoplasm of macrophages and microglia (Bagasra et al., 1995). Earlier histochemical studies identified reactive astrocytes as the major producers of nitric oxide (Bö et al., 1994). Macrophages/microglia could be labelled with an antibody recognizing nitrosylated proteins indicating concomitant reactive oxygen radical production with formation of peroxynitrite and protein nitrosylation (Bagasra et al., 1995).

Since work in EAE is controversial (Zielasek et al., 1995; Cross et al., 1994) the exact pathogenic role of nitric oxide remains to be clarified.

Complement constitutes the major effector mechanism of humoral immunity. Activated complement components are present in the lesions of EAE and MS (Compston et al., 1989, 1991). Increased concentrations were measured in plasma and cerebrospinal fluid of MS patients (Sanders et al., 1986). Administration of a recombinant complement regulatory protein, the soluble complement receptor 1 (CR1) was found to abrogate Lewis rat EAE (Piddlesden et al., 1994). Insertion of the membrane attack complex (C5b-9) into the myelin membrane opens transmembrane pores allowing for the influx of calcium. This disturbs the local ionic milieu impeding impulse propagation and activates myelin integral proteases to cause myelin breakdown.

Termination of the acute inflammatory episode

A number of mechanisms normally operate to counteract inflammation and stop the ongoing immune reponse.

Apoptosis of antigen-specific T cells in the lesions of EAE has been identified as an effective mechanism in stopping inflammation (Pender et al., 1992; Schmied et al., 1993). Programmed cell death of these infiltrating T cells might be the consequence of enhanced corticosteroid production, the presence or absence of cytokines, lack of availability of co-stimulatory molecules, or antigen induction (Bauer et al., 1995).

The disturbed balance between Th1 and Th2 responses may be restored through enhanced activity of Th2 cytokines. Indeed, increased levels of IL-10 and TGF-β mRNA expression in mononuclear cells from the peripheral blood and CSF are associated with the recovery phase of a relapse, periods of remission and a less aggressive disease course in MS (Söderström et al., 1995; Link et al., 1994; Rieckmann et al., 1994, 1995).

Early in the course of MS, these mechanisms may allow for complete recovery once the acute inflammatory episode abates. However, as the disease progresses, the capacity of the immune system to restore the disturbed balance between proinflammatory and downmodulating influences will be exhausted and a broadened autoaggressive response will inflict greater damage on the neural tissue.

References

Allegretta M, Nicklas JA, Sriram S, Albertini RJ (1990) T cells responsive to myelin basic protein in patients with multiple sclerosis. Science 247: 718–721

Bagasra O, Michaels FH, Zheng YM, Bobroski LE, Spitsin SV, Fu ZF, Tawadros R, Koprowski H (1995) Activation of the inducible form of nitric oxide synthase in the brains of patients with multiple sclerosis. Proc Natl Acad Sci USA 92: 12041–12045

Battistini L, Selmaj K, Kowal C, Ohmen J, Modlin RL, Raine CS, Brosnan CF (1995) Multiple sclerosis: limited diversity of the V2-J3 T cell receptor in chronic active lesions. Ann Neurol 37: 198–203

Bauer J, Wekerle H, Lassmann H (1995) Apoptosis in brain-specific autoimmune disease. Current Opinion in Immunology 7: 839–843

Beck J, Rondot P, Catinot L, Falcoff E, Kirchner H, Wietzerbin J (1988) Increased production of interferon-gamma and tumor necrosis factor precedes clinical manifestation in multiple sclerosis. Acta Neurol Scand 78: 318–323

Birnbaum G (1995) Stress proteins: their role in the normal central nervous system and in disease states, especially multiple sclerosis. Springer Seminars in Immunopathology 17: 107–118

Bö L, Dawson TM, Wesselingh S, Mörk S, Choi S, Kong PA, Hanley D, Trapp BD (1994) Induction of nitric oxide synthase in demyelinating regions of multiple sclerosis brains. Ann Neurol 36: 778–786

Butcher EC, Picker LJ (1996) Lymphocyte homing and homeostasis. Science 272: 60–66

Chia LS, Thompson JE, Moscarello MA (1983) Disorder in human myelin induced by superoxide radical: an in vitro investigation. Biochem Biophys Res Commun 117: 141–146

Chofflon M, Juillard C, Juillard P, Gauthier G, Grau GE (1992) Tumor necrosis factor alpha production as a possible predictor of relapse in patients with multiple sclerosis. Eur Cytokine Netw 3: 523–531

Compston DA, Morgan BP, Campbell AK, Wilkins P, Cole G, Thomas ND, Jasani B (1989) Immunocytochemical localization of the terminal complement complex in multiple sclerosis. Neuropathol Appl Neurobiol 15: 307–316

Compston DA, Scolding N, Noble M (1991) Pathogenesis of demyelinating diseases: Insights form cell biology. Trends Neurosci 14: 175–182

Cross AH, Misko TP, Lin RF, Hickey WF, Trotter JL, Tilton RG (1994) Aminoguanidine, an inhibitor of inducible nitric oxide synthase, ameliorates experimental autoimmune encephalomyelitis in SJL mice. J Clin Invest 93: 2684–2690

Cross AH, Girard TJ, Giacoletto KS, Evans RJ, Keeling RM, Lin RF, Trotter JL, Karr RW (1995) Long-term inhibition of murine experimental autoimmune encephalomyelitis using CTLA-4-Fc supports a key role for CD28 costimulation [see comments]. J Clin Invest 95: 2783–2789

De Simone R, Giampaolo A, Giometto B, Gallo P, Levi G, Peschle C, Aloisi F (1995) The costimulatory molecule B7 is expressed on human microglia in culture and in multiple sclerosis acute lesions. J Neuropathol Exp Neurol 54: 175–187

Freedman MS, Ruijs TC, Selin LK, Antel JP (1991) Peripheral blood gamma-delta T cells lyse fresh human brain-derived oligodendrocytes [see comments]. Ann Neurol 30: 794–800

Gijbels K, Masure S, Carton H, Opdenakker G (1992) Gelatenase in the cerebrospinal fluid of patients with multiple sclerosis and other inflammatory neurological disorders. J Neuroimmunol 41: 29–34

Gijbels K, Galardy RE, Steinman L (1994) Reversal of experimental autoimmune encephalomyelitis with a hydroxamate inhibitor of matrix metalloproteases. J Clin Invest 94: 2177–2182

Hartung HP, Archelos JJ, Zielasek J, Gold R, Koltzenburg M, Reiners K-H, Toyka KV (1995) Circulating adhesion molecules and inflammatory mediators in demyelination: A review. Neurology 45: 22–32

Hartung HP, Jung S, Stoll G, Zielasek J, Schmidt B, Archelos JJ, Toyka KV (1992) Inflammatory mediators in demyelinating disorders of the CNS and PNS. J Neuroimmunol 40: 197–210

Hartung HP, Michels M, Reiners K, Seeldrayers P, Archelos JJ, Toyka KV (1993) Soluble ICAM-1 serum levels in multiple sclerosis and viral encephalitis. Neurology 43: 2331–2335

Hartung HP, Reiners K, Archelos JJ, Michels M, Seeldrayers P, Heidenreich F, Pflughaupt KW, Toyka KV (1995b) Circulating adhesion molecules and tumor necrosis factor receptor in multiple sclerosis: correlation with magnetic resonance imaging. Ann Neurol 38: 186–193

Hauser SL, Doolittle TH, Lincoln R, Brown RH, Dinarello CA (1990) Cytokine accumulation in CSF of multiple sclerosis patients: frequent detection of interleukin-1 and tumor necrosis factor but not interleukin-6. Neurology 40: 1735–1739

Hofman FM, Hinton DR, Johnson K, Merrill JE (1989) Tumor necrosis factor identified in multiple sclerosis brain. J Exp Med 170: 607–612

Hohlfeld R, Meinl E, Weber F, Zipp F, Schmidt S, Sotgiu S, Goebels N, Voltz R, Spuler S, Iglesias A, et al (1995) The role of autoimmune T lymphocytes in the pathogenesis of multiple sclerosis. Neurology 45: 33–38

Konat GW, Offner H (1983) Density distribution of myelin fragments isolated from control and multiple sclerosis brain. Neurochem Int 4: 241–246

Konat GW, Wiggins RC (1985) Effect of reactive oxygen species on myelin membrane proteins. J Neurochem 45: 1113–1118

Kuchroo VK, Das MP, Brown JA, Ranger AM, Zamvil SS, Sobel RA, Weiner HL, Nabavi N, Gilmcher LH (1995) B7-1 and B7-2 costimulatory molecules activate differentially the Th1/Th2 developmental pathways: application to autoimmune disease therapy. Cell 80: 707–718

Kuroda Y, Shimamoto Y (1991) Human tumor necrosis factor-alpha augments experimental allergic encephalomyelitis in rats. J Neuroimmunol 34: 159–164

Lehmann PV, Sercarz EE, Forsthuber T, Dayan CM, Gammon G (1993) Determinant spreading and the dynamics of the autoimmune T-cell repertoire. Immunology Today 14: 203–208

Link J, Söderström M, Olsson T, Bo H, Ljungdahl A, Link H (1994) Increased transforming growth factor-β, interleukin-4, and interferon-τ in multiple sclerosis. Ann Neurol 36: 379–386

Markovic Plese S, Fukaura H, Zhang J, Al Sabbagh A, Southwood S, Sette A, Kuchroo VK, Hafler DA (1995) T cell recognition of immunodominant and cryptic proteolipid protein epitopes in humans. J Immunol 155: 982–992

Martino G, Moiola L, Brambilla E, Clementi E, Comi G, Grimaldi LME (1995) Interferon-gamma induces T lymphocyte proliferation in multiple sclerosis via a Ca^{2+}-dependent mechanism. J Neuroimmunol 62: 169–176

Matsuda M, Tsukada N, Miyagi K, Yanagisawa N (1994) Increased levels of soluble tumor necrosis factor receptor in patients with multiple sclerosis and HTLV-1-associated myelopathy. J Neuroimmunol 52: 33–40

Miller SD, McRae BL, Vanderlugt CL, Nikcevich KM, Pope JG, Pope L, Karpus WJ (1995) Evolution of the T-cell repertoire during the course of experimental immune-mediated demyelinating diseases. Immunol Rev 144: 225–244

Olsson T (1995) Critical influences of the cytokine orchestration on the outcome of myelin antigen-specific T-cell autoimmunity in experimental autoimmune encephalomyelitis and multiple sclerosis. Immunol Rev 144: 245–268

Olsson T, Wang W-Z, Höjeberg B, Kostulas V, Jiang Y-P, Anderson G, Ekre H-P, Link H (1990) Autoreactive T lymphocytes in multiple sclerosis determined by antigen-induced secretion of interferon-γ. J Clin Invest 86: 981–985

Panitch HS, Hirsch RL, Schindler J, Johnson KP (1987) Treatment of multiple sclerosis with gamma interferon: exacerbation associated with activation of the immune system. Neurology 37: 1097–1102

Pender MP, McCombe PA, Yoong G, Nguyen KB (1992) Apoptosis of alpha beta T lymphocytes in the nervous system in experimental autoimmune encephalomyelitis: its possible implications for recovery and acquired tolerance. J Autoimmun 5: 401–410

Piddlesden SJ, Storch MK, Hibbs M, Freeman AM, Lassmann H, Morgan BP (1994) Soluble recombinant complement receptor 1 inhibits inflammation and demyelination in antibody-mediated demyelinating experimental allergic encephalomyelitis. J Immunol 152: 5477–5484

Racke MK, Scott DE, Quigley L, Gray GS, Abe R, June CH, Perrin PJ (1995) Distinct roles for B7-1 (CD80) and B7-2 (CD86) in the initiation of experimental allergic encephalomyelitis. J Clin Invest 96: 2195–2203

Raine CS (1994) The Dale E, McFarlin Memorial Lecture: the immunology of the multiple sclerosis lesion. Ann Neurol 36 [Suppl]: 61–72

Rieckmann P, Albrecht M, Kitze B, Weber T, Tumani H, Broocks A, Lüer W, Poser S (1994a) Cytokine mRNA levels in mononuclear blood cells from patients with multiple sclerosis. Neurology 44: 1523–1526

Rieckmann P, Martin S, Weichselbraun I, Albrecht M, Kitze B, Weber T, Tumani H, Broocks A, Lüer W, Helwig A, Poser S (1994b) Serial analysis of circulating adhesion molecules and TNF receptor in serum from patients with multiple sclerosis.: clCAM-1 is an indicator for relapse. Neurology 44: 2367–2372

Rieckmann P, Albrecht M, Kitze B, Weber T, Tumani H, Broocks A, Lüer W, Helwig A, Poser S (1995) Tumor necrosis factor-alpha messenger RNA expression in patients with relapsing-remitting multiple sclerosis is associated with disease activity. Ann Neurol 37: 82–88

Rosenberg GA, Dencoff JE, Correa N, Reiners M, Ford CC (1996) Effect of steroids on CSF matrix metalloproteinases in multiple scelerosis: relation to blood-brain barrier injury. Neurology 46: 1626–1632

Ruddle NH, Bergman CM, McGrath KM, Lingenheld EG, Grunnet ML, Padula SJ, Clark RB (1990) An antibody to lymphotoxin and tumor necrosis factor prevents transfer of experimental allergic encephalomyelitis. J Exp Med 172: 1193–1200

Sanders ME, Koski CL, Robbins D, Shin ML, Frank MM, Joiner KA (1986) Activated terminal complement in cerebrospinal fluid in Guillain-Barre syndrome and multiple sclerosis. J Immunol 136: 4456–4459

Schmied M, Breitschopf H, Gold R, Zischler H, Rothe G, Wekerle H, Lassmann H (1993) Apoptosis of T lymphocytes in experimental autoimmune encephalomyelitis. Evidence for programmed cell death as a mechanism to control inflammation in the brain. Am J Pathol 143: 446–452

Selmaj K, Raine CS (1988) Tumor necrosis factor mediates myelin and oligodendrocyte damage in vitro. Ann Neurol 23: 339–346

Selmaj K, Brosnan CF, Raine CS (1991a) Colocalization of lymphocytes bearing gamma delta T-cell receptor and heat shock protein hsp65+ oligodendrocytes in multiple sclerosis. Proc Natl Acad Sci USA 88: 6452–6456

Selmaj K, Raine CS, Cannella B, Brosnan CF (1991b) Identification of lymphotoxin and tumor necrosis factor in multiple sclerosis lesions. J Clin Invest 87: 949–954

Selmaj K, Raine CS, Cross AH (1991c) Anti-tumor necrosis factor therapy abrogates autoimmune demyelination. Ann Neurol 30: 694–700

Sharief MK, Hentges R (1991) Association between tumor necrosis factor-α and disease progression in patients with multiple sclerosis. N Engl J Med 325: 467–472

Sharief MK, Noori MA, Ciardi M, et al (1993) Increased levels of circulating ICAM-1 in serum and cerebrospinal fluid of patients with active multiple sclerosis: correlation with TNF-α and blood-brain barrier damage. J Neuroimmunol 43: 15–22

Sippy BD, Hofman FM, Wallach D, Hinton DR (1995) Increased expression of tumor necrosis factor-alpha receptors in the brains of patients with AIDS. J-Acquir Immune Defic Syndr Hum Retrovirol 10: 511–521

Söderström M, Hillert J, Link J, Navikas V, Fredrikson S, Link H (1995) Expression of

IFN-gamma, IL-4 and TGF-beta in multiple sclerosis in relation to HLA-Dw2 phenotype and stage of disease. Multiple Sclerosis 1: 173–180

Steinman L, Waisman A, Altmann D (1995) Major T-cell responses in multiple sclerosis. Mol Med Today 1: 79–83

Stinissen P, Vandevyver C, Medaer R, Vandegaer L, Nies J, Tuyls L, Hafler DA, Raus J, Zhang J (1995) Increased frequency of gamma delta T cells in cerebrospinal fluid and peripheral blood of patients with multiple sclerosis. Reactivity, cytotoxicity, and T cell receptor V gene rearrangements. J Immunol 154: 4883–4894

Tchélingérian JL, Monge M, Le Saux F, Zalc B, Jacque C (1995) Differential oligodendroglial expression of the tumor necrosis factor receptors in vivo and in vitro. J Neurochem 65: 2377–2380

Theofilopoulos AN (1995) The basis of autoimmunity. Part I. Mechanisms of aberrant self-recognition. Immunol Today 16: 90–98

Tracey KJ, Cerami A (1993) Tumor necrosis factor: an updated review of its biology. Crit Care Med 21: 415–422

Trotter JL, Collins KG, Van der Veen RC (1991) Serum cytokine levels in chronic progressive multiple sclerosis: interleukin-2 levels parallel tumor necrosis factor-alpha levels. J Neuroimmunol 33: 29–36

Tsukada N, Miymgi K, Matsuda M, et al (1993) Increased levels of circulating intercellular adhesion molecule-1 in multiple sclerosis and human T-lymphotropic virus type 1-associated myelopathy. Ann Neurol 33: 591–596

Van Noort JM, Van Sechel AC, Bajramovic JJ, El Ouagmiri M, Polman CH, Lassmann H, Ravid R (1995) The small heat-shock protein alpha B-crystallin as candidate autoantigen in multiple sclerosis. Nature 375: 798–801

Voskuhl RR, Martin R, Bergman C, Dalal M, Ruddle NH, McFarland HF (1993) T helper (th1) functional phenotype of human myelin basic protein-specific T lymphocytes. Autoimmunity 15: 137–143

Warren KG, Catz I, Steinman L (1995) Fine specificity of the antibody response to myelin basic protein in the central nervous system in multiple sclerosis: the minimal B-cell epitope and a model of its features. Proc Natl Acad Sci USA 92: 11061–11065

Windhagen A, Newcombe J, Dangond F, Strand C, Woodroofe MN, Cuzner ML, Hafler DA (1995) Expression of costimulatory molecules B7-1 (CD80), B7-2 (CD86), and interleukin 12 cytokine in multiple sclerosis lesions. J Exp Med 182: 1985–1996

Wucherpfennig KW, Hafler DA (1995) A review of T-cell receptors in multiple sclerosis: clonal expansion and persistence of human T-cells specific for an immunodominant myelin basic protein peptide. Ann NY Acad Sci 756: 241–258

Wucherpfennig KW, Strominger JL (1995) Molecular mimicry in T cell-mediated autoimmunity: viral peptides activate human T cell clones specific for myelin basic protein. Cell 80: 695–705

Wucherpfennig KW, Newcombe J, Li H, Keddy C, Cuzner ML, Hafler DA (1992) Gamma delta T-cell receptor repertoire in acute multiple sclerosis lesions. Proc Natl Acad Sci USA 89: 4588–4592

Wucherpfennig KW, Zhang J, Witek C, Matsui M, Modabber Y, Ota K, Hafler DA (1994) clonal expansion and persistence of human T cells specific for an immunodominant myelin basic protein peptide. J Immunol 152: 5581–5592

Zielasek J, Jung S, Gold R, Liew FY, Toyka KV, Hartung HP (1995) Administration of nitric oxide synthase inhibitors in experimental autoimmune neuritis and experimental autoimmune encephalomyelitis. J Neuroimmunol 58: 81–88

Zipp F, Weber F, Huber S, Sotgiu S, Czlonkowska A, Holler E, Albert E, Weiss EH, Wekerle H, Hohlfeld R (1995) Genetic control of multiple sclerosis: increased production of lymphotoxin and tumor necrosis factor-alpha by HLA.-DR2[+] T cells. Ann Neurol 38: 723–730

Authors' address: Prof. H.-P. Hartung, Neurologische Universitätsklinik, Josef-Schneider-Strasse 11, D-97080 Würzburg, Federal Republic of Germany.

Basic mechanisms of brain inflammation

H. Lassmann

Clinical Institute of Neurology, University of Vienna, Vienna, Austria

Summary. The mechanisms, how the immune system surveys the nervous tissue and how brain inflammation is regulated are essential questions for therapy of neuroimmunological diseases. The nervous system is continuously patrolled by hematogenous cells, which may pass the blood brain barrier in an activated state. When these cells find their respective target antigen in the CNS compartment, an inflammatory reaction is started through the secretion of proinflammatory cytokines. This leads to the upregulation of endothelial adhesion molecules and the local production of chemokines, which in concert facilitate the entry of inflammatory effector cells into the lesions. T-lymphocytes are effectively removed from inflammatory brain lesions by local apoptosis. In addition some lymphatic drainage of the nervous system allows the removal of effector cells from the lesions and their migration into regional lymph nodes. In summary these data suggest that the immune surveillance of the central nervous system is much more tightly controlled compared to that in other organs.

Introduction

Inflammation in the central nervous system is instrumental in a variety of human diseases, such as acute meningitis or encephalitis as well as multiple sclerosis or AIDS. To understand the pathogenesis of inflammatory brain diseases several basic questions have to be addressed. The immune system has to recognize the pathogen or autoantigen, that is sequestered in the nervous system. This implies, that immune cells have to get access to the normal nervous system and that the antigens have to be presented in a way that allows recognition. When antigen is recognized a cascade of events has to be started which leads to inflammation and elimination of the pathogen and which may be associated with additional tissue damage. Finally, during clearance of inflammation, inflammatory cells must be removed from the lesions. Since the regeneration capacity of the nervous system is very limited, all these processes have to be accomplished in a way, that keeps unspecific tissue damage at the lowest possible level. One of the central questions in neuroimmunology is thus to elucidate, how the immune system accomplishes this task.

Immune privilege of the normal brain

In several immunological aspects the central nervous system differs from other organs. Due to the blood brain barrier immune cells, antibodies and immunological mediators have little access to the normal brain (Hickey et al., 1991). Yet, there are some hematogenous cells, in particular monocytes/macrophages and T-lymphocytes, present in the CNS tissue even under completely normal conditions (Hauser et al., 1983). Studies in bone marrow chimeras revealed that there is continuos turnover and replacement of monocytes and T-lymphocytes, mainly in the meninges and the perivascular space (Hickey and Kimura 1988; Hickey et al., 1992). Whereas at present it is not clear, how monocytes enter the CNS compartment, T-lymphocytes can traverse the normal blood brain barrier, when they are in a activated stage (Wekerle et al., 1986; Hickey et al., 1991). These data suggest, that it is the small pool of peripherally activated T-cells, that enter the CNS compartment for immune surveillance and are, thus, allowed to search for specific antigens or pathogens. Yet, when there is no peripheral T-cell activation, the respective antigens may hide in the CNS and escape detection by the immune system. Thus transplants may survive in the brain environment, in spite of a mismatch in histocompatibility antigens (Sloan et al., 1991; Broadwell et al., 1994), and by the same mechanism the brain protects itself against autoimmune reactions.

Another mechanism is the active suppression of the expression of major histocompatibility (MHC) antigens in the nervous system. Antigen recognition by T-lymphocytes requires the presentation of small peptides of the antigenic protein by antigen presenting cells in the context of MHC antigens. Thus, in the absence of MHC-antigens T-lymphocytes cannot recognize their target and, thus, do not activate the mechanisms, that finally lead to inflammation. MHC-expression in the normal brain is very low or even absent (Vass et al., 1986; Hart and Fabry, 1995). Its absence apparently does not reflect the lack of immunstimulatory signals, but is rather due to active suppression by electrically active nerve cells (Neumann et al., 1995). Damage of the nervous system, that blocks electrical activity, may induce MHC expression in neurons and possibly in neighboring glia cells. This allows T-cells to recognize their antigen, to induce inflammation or to eliminate their target through direct cytotoxicity.

The low expression of immune activation associated cell surface molecules in the CNS is not restricted to MHC-antigens. The expression of adhesion molecules, that are essential in cell-cell contacts during the migration of inflammatory cells (Bevilaqua, 1993; Springer, 1994), is low on endothelial cells of cerebral vessels as well as on local tissue elements (Sobel et al., 1990; Raine et al., 1990; Male et al., 1990; Lassmann et al., 1991; Rössler et al., 1992). Low expression is also found for CD 34, the receptor for lipopolysaccharide (LPS). Thus, induction inflammation by bacteria in the CNS, which is primarily mediated by LPS, is delayed in the brain and requires a very high bacterial load in comparison to that in peripheral organs (Quagliarello and Scheld, 1992; Lawson and Perry, 1995).

Finally the CNS is equipped with a particular tissue macrophage, the microglia (Germann et al., 1995). In the normal brain microglia in its resting state seems to be immunologically inert. Although microglia can become activated by a variety of immunological and non-immunological stimuli, different levels of activation appear to exist, which may transform the cells in either pure scavengers or fully activated immunological effector cells. Furthermore these cell can produce both, immunostimulatory as well as immunosuppressive cytokines (Kiefer et al., 1995). Thus it is still unclear, whether the prime role of microglia in brain inflammation is to propagate or suppress the immune response.

Immune surveillance and the induction of initial brain inflammation

As mentioned before, activated T-lymphocytes can pass the normal blood brain barrier. Thus, peripheral stimulation of the immune system, for instance in the course of an infection, will allow a selected population of T-cells to enter the CNS and to search for their respective antigen. When the antigen or a cross reactive autoantigen is present in the CNS compartment, it will be presented to the T-lymphocytes on a population of meningeal and perivascular tissue macrophages, which constitutively express MHC antigens even under normal conditions (Vass et al., 1986; Hickey and Kimura, 1988). The initiation and propagation of brain inflammation can be mediated by antigen recognition on the tissue macrophages in the meninges and the perivascular space alone and does not require the participation of other cells of the CNS parenchyme, such as microglia and astrocytes (Hickey and Kimura, 1988; Lassmann et al., 1993).

The initial antigen recognition in the perivascular space triggers a cascade of secondary events, that lead to the development of brain inflammation. A key role in this process is played by cytokines, that are produced either by hematogenous cells or by local tissue elements. In acute T-cell mediated encephalomyelitis a characteristic temporal sequence of cytokine production has been observed. An initial peak of Interleukin 12 and gamma-interferon production is rapidly followed by pronounced synthesis of Interleukin 1 and Tumor Necrosis Factor alpha (Olsson, 1994). The latter correlates well with the peak of tissue infiltration with hematogenous cells. This process is associated with a pronounced upregulation of the expression of adhesion molecules on the endothelial cells of cerebral vessels (Lassmann et al., 1991; Cannella and Raine, 1995). The endothelial adhesion molecules apparently are required in the recruitment of secondary, non activated hematogenous effector cells, which are instrumental in the removal of pathogens but also in the induction of inflammatory tissue damage.

Amplification of the inflammatory response by local production of chemokines

Chemokines are small peptides, which are chemotactic for other inflammatory cells and play an important role in the secondary recruitment of leukocytes into an established inflammatory focus (Baggiolini et al., 1994). In brain inflammation chemokines are not only produced by hematogenous cells but also and prominently by local cells such as astrocytes (Tani and Ransohoff, 1994). Liberated into the extracellular and perivascular space, they apparently modulate the inflammatory reaction by attracting additional monocytes or – in the case of bacterial meningitis – granulocytes into the lesions. In comparison to cytokines, the synthesis of chemokines in the lesions is much more prominent and less tightly controlled (Schlüsener and Meyermann, 1993). It has, thus, been suggested that chemokines are instrumental in the amplification of the inflammatory reaction and in the spread of inflammation into the parenchyme of the CNS.

Recently, however, it became clear, that not only the classical chemokines may exert chemotactic functions. Certain neuropeptides, as for instance Substance P, Vascular Intestinal Peptide or Secretoneurin, have been shown to be leucotactic for T-cells, monocytes or granulocytes (Carolan and Casale, 1993; Reinisch et al., 1993; Johnston et al., 1994). Interestingly, in acute T-cell mediated brain inflammation, a significant association between local expression of Secretoneurin with macrophage infiltration has been observed (Storch et al., 1996). This suggest, that in addition to classical chemokines the local milieu of neurotransmitters may modulate inflammation in the brain.

Mechanisms of immune-mediated tissue damage in inflammatory brain lesions

Immune effector cells, such as monocytes/macrophages, granulocytes or cytotoxic T-cells produce a variety of toxic factors, that are required for the destruction and elimination of foreign pathogens. These include cytotoxic cytokines (e.g. TNF-alpha, lymphotoxin), perforin or complement components, proteolytic and lipolytic enzymes and oxygen radicals. These toxic factors, however, not only destroy foreign pathogens but may also damage local tissue elements. In the central nervous system the myelin/oligodendroglia complex appears to be particularly vulnerable to the action of these toxic inflammatory mediators (Griot et al., 1990; Scolding et al., 1990; Selmaj and Raine, 1988). In vitro, in mixed culture systems, it is generally the oligodendrocyte with its myelin sheaths, that is damaged to a much larger extent as other CNS elements, such as astrocytes or neurons. This may in part explain, why chronic inflammatory conditions of the central nervous system in experimental autoimmune encephalomyelitis as well as in human conditions, as multiple sclerosis or HIV-encephalitis are accompanied by extensive white matter pathology.

In addition, however, the inflammatory response may directly affect neurons. Activated macrophages can produce significant amounts of quinolinic acid and other excitotoxins, which – at least in vitro – may destroy nerve cells through an NMDA-receptor mediated pathway (Giulian et al., 1993; Lipton et al., 1994; Espey et al., 1995). This mechanism may play a significant role in the induction of neuronal loss in HIV-encephalitis and may also contribute to axonal damage in multiple sclerosis lesions.

Clearance of inflammation

Inflammatory cells, that have entered the brain during inflammation, have to be removed from the lesions during recovery. In contrast to the mechanisms, that are involved in the induction of brain inflammation, those that operate during clearance are much less understood. Pender et al. (1991) first reported that in T-cell mediated inflammatory lesions of the brain abundant cells are locally destroyed by apoptosis. Later it became clear, that most of the apoptotic cells are T-lymphocytes and that the peak of T-cell apotosis correlates well with the clearance of the inflammatory lesions (Schmied et al., 1993). Both, direct as well as indirect evidence suggests, that it is the population of autoantigen-specific T-lymphocytes that are removed by programmed cell death in the brain (Tabi et al., 1994; Bauer et al., 1995). A similar destruction of T-cells by apoptosis has been identified in autoimmune neuritis (Zettel et al., 1994), corona virus induced demyelinating encephalomyelitis (Barac-Latas et al., 1995) and multiple sclerosis (Ozawa et al., 1995), but not in inflammatory diseases of peripheral organs (Bauer et al., 1995). Therefore, it appears that the nervous system has established a specific way to downregulate T-cell mediated immune responses through the local destruction of antigen-specific T-cells. The mechanisms of apoptosis induction in these conditions are still poorly understood.

Although apoptosis is sometimes also encountered in macrophages (Nguyen et al., 1994) and granulocytes, its incidence in inflammatory brain lesions is too low to explain their removal by this mechanism. Therefore some drainage of secondary effector cells from the brain into the blood vessels or the local lymphatic tissue has to be considered. The existence of lymphatic drainage pathways in the nervous system has for long been controversial. Recently, however, it has been shown that antigens or particulate material, that were injected into the cerebrospinal fluid may reach regional lymph systems at areas, where cranial and spinal nerves and blood vessels pass the meningeal barrier (Weller et al., 1992; Zenker et al., 1994). Such intrathecally injected antigens may elicit a local immune response for instance in the deep cervical lymph nodes (Cserr et al., 1992). We have recently studied the migration routes of macrophages, that have taken up brain proteins in inflammatory demyelinating lesions and found evidence for a drainage pathway through the spinal meninges into the epidural lymphatic vessels and the paraaortal lymph nodes. In addition, some macrophages were also found to migrate through the walls of inflamed vessels into the circulation and could then be identified in the spleen. Thus, these data suggest that there is indeed a lymphatic drainage of the central nervous system. In contrast to lymphatic drainage in other organs, however, the time, required for macrophages to reach the lymph nodes was much longer. Thus, the transport of antigens from the CNS into the lymphatic tissue may be limited by the fact, that most antigens will already be degraded when the macrophages accumulate in the lymphatic environment.

Conclusions

In summary the mechanisms of brain inflammation in many respects follow the same basic patterns, that operate in inflammation at other sites of the body. Yet, the brain has developed a variety of protective mechanisms, which secure an efficient immune surveillance while keeping undesired immune mediated tissue damage at a lowest possible level. These protective mechanisms are partly accomplished by the blood brain barrier, that allows the entry of immune effector cells, immune mediators and antibodies only when damage already has occurred. Furthermore, the low expression of histocompatibility antigens and other immune associated cell surface receptors and the efficient elimination of antigen-specific T-lymphocytes from the brain in general prevents overheated local immune responses, that could be deleterious for brain function. Yet, when very severe proinflammatory stimuli are provided the full immunological repertoire of inflammation can be activated. Thus, in the CNS inflammation is a graded response, that secures an immune surveillance at the lowest level, that is just appropriate for the defense against pathogens. Obviously, a dysregulation of this process, which may take place in certain infectious diseases or in autoimmunity has deleterious effects on brain integrity.

Acknowledgement

This study was partly funded by Austrian Science Foundation Project P10608 MED.

References

Baggiolini M, Dewald B, Moser B (1994) Interleukin 8 and related chemotactic cytokines – CXC and CC chemokines. Adv Immunol 55: 97–179

Barac-Latas V, Wege H, Lassmann H (1995) Apoptosis of T lymphocytes in coronavirus induced encephalomyelitis. Regional Immunol 6: 355–357

Bauer J, Wekerle H, Lassmann H (1996) Apoptosis in brain-specific autoimmune disease. Curr Opinion Immunol 7: 839–843

Bevilaqua MP (1993) Endothelial-leucocyte interactions and regulation of leucocyte migration. Ann Rev Immunol 11: 767–804

Broadwell RD, Baker BJ, Ebert PS, Hickey WF (1994) Allografts of CNS tissue possess a blood brain barrier: III. Neuropathological, methodological, and immunological considerations. Microscopy Res Tech 27: 471–494

Canella B, Raine CS (1995) The adhesion molecule and cytokine profile of multiple sclerosis lesions. Ann Neurol 37: 424–435

Carolan EJ, Casale TB (1993) Effects of neuropeptides on neutrophil migration through noncellular and cellular endothelial barriers. J Allergy Clin Immunol 92: 589–598

Cserr HF, Harling-Berg CJ, Knopf PM (1992) Drainage of brain extracellular fluid into blood and deep cervical lymph and its immunological significance. Brain Pathol 2: 269–276

Espey MG, Moffett JR, Narnboodiri MA (1995) Temporal and spatial changes of quinolinic acid immunoreactivity in the immune system of lipopolysaccharide-stimulated mice. J Leukoc Biol 57: 199–206

Gehrmann J, Matsumoto Y, Kreutzberg GW (1995) Microglia: intrinsic immuneffector cell of the brain. Brain Res Rev 20: 269–287

Giulian D, Wendt E, Vaca K, Noonan CA (1993) The envelope glycoprotein of human immunodeficiency virus type-1 stimulate and release neurotoxins from monocytes. Proc Natl Acad Sci (USA) 90: 2769–2773

Griot C, Vandevelde M, Richard A, Peterhans E, Stocker R (1990) Selective degeneration of oligodendrocytes mediated by reactive oxygen species. Free Radic Res Commun 11: 181–193

Hart MN, Fabry Z (1995) CNS antigen presentation. Trends Neurosci 18: 475–481

Hauser SL, Bhan AK, Gilles F, Hoban CJ, Reinherz EL, Schlossman SF, Weiner HL (1983) Immunohistochemical staining of human brain with monoclonal antibodies that identify lymphocytes, monocytes and the Ia-antigen. J Neuroimmunol 5: 197–205

Hickey WF, Hsu BL, Kimura H (1991) T-lymphocyte entry in the central nervous system. J Neurosci Res 28: 254–260

Hickey WF, Kimura H (1988) Perivascular microglia cells of the CNS are bone marrow derived and present antigen in vivo. Science 239: 290–292

Hickey WF, Vass K, Lassmann H (1992) Bone marrow derived elements in the central nervous system: An immunohistochemical and ultrastructural survey of rat chimeras. J Neuropath Exp Neurol 51: 246–256

Johnston JA, Taub DD, Lloyd AR, Conlon K, Oppenheim JJ, Kevlin DJ (1994) Human T-lymphocyte chemotaxis and adhesion induced by vasoactive intestinal peptide. J Immunol 153: 1762

Kiefer R, Streit WJ, Toyka KV, Kreutzberg GW, Hartung HP (1995) Transforming growth factor beta-1: a lesion-associated cytokine of the nervous system. Int J Dev Neurosci 13: 331–339

Lassmann H, Rössler K, Zimprich F, Vass K (1991) Expression of adhesion molecules and histocompatibility antigens at the blood-brain barrier. Brain Pathol 1: 115–123

Lassmann H, Schmied M, Vass K, Hickey WF (1993) Bone marrow derived elements and resident microglia in brain inflammation. Glia 7: 19–24

Lawson LJ, Perry VH (1995) The unique characteristics of inflammatory responses in mouse brain are acquired during postnatal development. Eur J Neurosci 7: 1584–1595

Lipton SA, Yeh M, Dreyer EB (1994) Update on current models of HIV-related neuronal injury: platelet-activating factor, arachidonic acid and nitric oxide. Adv Neuroimmunol 4: 181–188

Male D, Pryce G, Rahman J (1990) Comparison of the immunological properties of rat cerebral and aortic endothelium. J Neuroimmunol 30: 161–168

Neumann H, Cavalle A, Jenne DE, Wekerle H (1995) Induction of MHC I genes in neurons. Science 269: 549–551

Nguyen KB, McCombe PA, Pender MP (1994) Macrophage apoptosis in the central nervous system in experimental autoimmune encephalomyelitis. J Autoimmun 7: 145–152

Olsson T (1994) Role of cytokines in multiple sclerosis and experimental autoimmune encephalomyelitis. Eur J Neurol 1: 7–19

Ozawa K, Suchanek G, Breitschopf H, Bruck W, Budka H, Jellinger K, Lassmann H (1994) Patterns of oligodendroglia pathology in multiple sclerosis. Brain 117: 1311–1322

Pender MP, Nguyen KB, McCombe PA, Kerr JFR (1991) Apoptosis in the nervous system in experimental allergic encephalomyelitis. J Neurol Sci 104: 81–87

Quagliarello V, Scheld WM (1992) Bacterial meningitis: pathogenesis, pathophysiology and progress. N Engl J Med 327: 864–872

Raine CS, Lee SC, Scheinberg LC, Duijvestin AM, Cross AH (1990) Adhesion molecules on endothelial cells in the central nervous system: an emerging area in the neuroimmunology of multiple sclerosis. Clin Immunol Immunopathol 57: 173–187

Reinisch N, Kirchmair R, Kahler CM, Hogue-Angeletti R, Fischer-Colbrie R, Winkler H, Wiedermann CJ (1993) Attraction of human monocytes by the neuropeptide secretoneurin. FEBS Lett 334: 41–44

Rössler K, Neuchrist C, Kitz K, Scheiner O, Kraft D, Lassmann H (1992) Expression of leucocyte adhesion molecules at the human blood-brain-barrier. J Neurosci Res 31: 365–374

Schlüsener HJ, Meyermann R (1993) Intercrines in brain pathology. Expression of intercrines in a multiple sclerosis and a Creutzfeld-Jacob lesion. Acta Neuropathol 86: 393–396

Schmied M, Breitschopf H, Gold R, Zischler H, Rothe G, Wekerle H, Lassmann H (1993) Apoptosis of T lymphocytes in experimental autoimmune encephalomyelitis: evidence for programmed cell death as a mechanism to control inflammation in the brain. Am J Pathol 143: 446–452

Scolding N, Jones J, Compston DA, Morgan BP (1990) Oligodendrocyte susceptibility to injury by T-cell perforin. Immunology 70: 6–10

Selmaj KW, Raine CS (1988) Tumor necrosis factor mediates myelin and oligodendrocyte damage in vitro. Ann Neurol 23: 339–346

Sloan DJ, Wood MJ, Charlton HM (1991) The immune response to intracerebral neural grafts. Trends Neurosci 14: 341–346

Sobel RA, Mitchell ME, Fondren G (1990) Intercellular adhesion molecule-1 (ICAM-1) in cellular immune reactions in the human central nervous system. Am J Pathol 136: 1309–1316

Springer TA (1994) Traffic signals for leucocyte emigration: the multistep paradigm. Cell 76: 301–314

Storch MK, Fischer Colbrie R, Smith T, Rinner WA, Hickey WF, Cuzner ML, Winkler H, Lassmann H (1996) Colocalization of secretoneurin immunoreactivity and macrophage infiltration in the lesions of experimental autoimmune encephalomyelitis. Neuroscience (in press)

Tabi Z, McCombe PA, Pender MP (1994) Apoptotic elimination of Vβ8.2+ cells form the central nervous system during recovery from experimental autoimmune encephalomyelitis induced by passive transfer of Vβ8.2 encephalitogenic T-cells. Eur J Immunol 24: 2609–2617

Tani M, Ransohoff RM (1994) Do chemokines mediate inflammatory cell invasion of the central nervous system parenchyma? Brain Pathol 4: 135–143

Vass K, Lassmann H, Wekerle H, Wisniewski HM (1986) The distribution of Ia-antigen in the lesions of rat acute experimental allergic encephalomyelitis. Acta Neuropathol 70: 149–160

Wekerle H, Linington C, Lassmann H, Meyermann R (1986) Cellular immune reactivity within the CNS. Trends Neurosci 9: 271–277

Weller RO, Kida S, Zhang ET (1992) Pathways of fluid drainage from the brain – morphological aspects and immunological significance in rat and man. Brain Pathol 2: 277–284

Zenker W, Bankoul S, Braun JS (1994) Morphological indications for considerable diffues reabsorption of cerebrospinal fluid in spinal meninges particularly in the areas of meningeal funnels. An electronmicroscopical study including tracing experiments in rats. Anat Embryol (Berl) 189: 243–258

Zettel U, Gold R, Hartung HP, Toyka KV (1994) Apoptotic cell death of T-lymphocytes in experimental autoimmune neuritis of the Lewis rat. Neurosci Lett 176: 75–79

Author's address: Prof. Dr. H. Lassmann, Institute of Neurology, University of Vienna, Schwarzspanierstrasse 17, A-1090 Wien, Austria.

Cell death in prion disease

H. A. Kretzschmar[1], A. Giese[1], D. R. Brown[1], J. Herms[1], B. Keller[2], B. Schmidt[3], and M. Groschup[4]

[1]Institut für Neuropathologie, [2]Abteilung Neuro- und Sinnesphysiologie, Zentrum Physiologie und Pathophysiologie, [3]Abteilung Biochemie II, Zentrum Biochemie und Molekulare Zellbiologie, Universität Göttingen, Göttingen, [4]Bundesanstalt für Viruskrankheiten der Tiere, Tübingen, Federal Republic of Germany

Summary. Prion diseases are neurodegenerative transmissible diseases. The infectious agent, termed prion, is thought to consist of an altered host-encoded protein. The pathogenesis of these diseases which typically in a very short time lead to rampant nerve cell death and astrocytic gliosis is poorly understood. Investigations using the in situ endlabeling technique and electron microscopy in a scrapie model in the mouse (79A strain) show that nerve cell death is due to apoptosis. A cell culture model using a synthetic peptide of the prion protein (PrP106-126) shows that this peptide is toxic only to normal neurons whereas nerve cells derived from PrP knock-out (PrP$^{0/0}$) mice are unaffected by this neurotoxic effect. In addition, microglia play a crucial part in this process by secreting reactive oxygen species. Experiments in animals will have to show whether these cell culture findings adequately reflect the in vivo pathogenesis.

Introduction

Prion diseases (transmissible spongiform encephalopathies) are a group of fatal progressive neurological disorders including Creutzfeldt-Jakob disease, Gerstmann-Sträussler-Scheinker syndrome, fatal familial insomnia and kuru in humans as well as bovine spongiform encephalopathy (BSE) and scrapie in animals. They are unique in being both horizontally transmissible as well as sometimes hereditary diseases (Prusiner, 1993). The infectious agent of these diseases has been termed prion (Prusiner, 1982). The only molecule consistently found in infectious preparations is the scrapie isoform of the prion protein (PrP^{Sc}). This protein derives from a normal cellular isoform of the prion protein (PrP^C) which is encoded in the genome of all mammals investigated to date (Prusiner, 1991). PrP^C is a glycoprotein of unknown function normally found in neurons (Kretzschmar et al., 1986) and glia (Moser et al., 1995).

The classical pathological features of spongiform encephalopathies include vacuolation of the neuropil, astrogliosis and neuronal loss (Masters and

Richardson, 1978). Alfons Jakob's description of "spastic pseudosclerosis-encephalomyelopathy" (Jakob, 1921) [later termed "Creutzfeldt-Jakob disease" (Spielmeyer, 1922)] did not include vacuolation or spongiform change as a pathological feature, although he mentions it later (Jakob, 1923). Neuronal degeneration is widespread in many cases of CJD and may lead to almost complete loss of nerve cells (Fig. 1A, B). Jakob describes a variety of pathological changes including chromatolysis of Betz cells in the motor cortex as well as satellitosis and neuronophagia ("glial rosettes"). However, there is only limited data available on the extent and pattern of neuronal loss (Masters and Richardson, 1978; Hogan et al., 1981; Scott and Fraser, 1984; Jeffrey et al., 1992, 1995), and even less is known about the mechanisms which underlie neuronal cell death in spongiform encephalopathies.

Cell culture experiments with neurotoxic prion protein fragments suggest that neuronal cell death in these diseases may be due to apoptosis. To test this hypothesis in vivo we used the in situ end-labeling (ISEL) technique and electron microscopy to study cell death in an experimental scrapie system in the mouse. ISEL, which relies on the incorporation of labeled nucleotides in fragmented DNA by terminal transferase, showed labeled nuclei in the brains and retinae of mice infected with the 79A strain of scrapie, whereas no labeling was observed in control animals. In the brain, labeled nuclei were mainly found in the granule layer of the cerebellum of terminally ill mice. These results support the hypothesis that neuronal loss in spongiform encephalopathies is due to apoptosis.

It is not clear whether apoptotic nerve cell loss is caused by progressive loss of the function of PrP^C or increasing toxicity of PrP^{Sc} ("gain of function hypothesis") in the course of the disease. Previous experiments have suggested that the normal cellular prion protein (PrP^C) is involved in synaptic function in the hippocampus (Collinge et al., 1994). This finding has been interpreted as an argument in favor of the "loss of function hypothesis". To investigate the synaptic function of prion protein in cerebellar Purkinje cells we utilized the controlled recording conditions of the patch clamp technique. In PrP gene-ablated mice ($PrP^{0/0}$ mice) (Büeler et al., 1992), the kinetics of GABA- and glutamate receptor-mediated currents showed no significant deviation from those in control animals. In consequence, our findings do not confirm the hypothesis that progressive loss of synaptic function of PrP^C causes neuronal apoptosis. Thus, the functional role of PrP^C remains enigmatic.

A fragment of human PrP which consists of amino acids 106-126 and forms fibrils in vitro has been previously demonstrated to be toxic to cultured hippocampal neurones (Forloni et al., 1993). In cell culture experiments we were able to demonstrate that PrP106-126 is toxic to cells from the cerebellum but that this toxicity requires the presence of microglia. In response to the presence of PrP106-126, microglia increase production of oxygen radicals. However, the effect of PrP106-126 on microglia, necessary for the toxic effect on normal neurones, is not sufficient to kill neurones from mice not expressing PrP^C. These findings are in accord with reports on the resistance of $PrP^{0/0}$ mice to scrapie. Our results are also indicative of a neurodegenerative mechanism similar to that of in vitro models of Alzheimer's disease.

1. Neuronal cell death in prion diseases is due to apoptosis

It is generally believed that there are two basic types and mechanisms of cell death: necrosis and apoptosis (Kerr et al., 1972; Searle et al., 1982; Buja et al., 1993). Necrosis often results from severe and sudden injury and leads to rapid cell lysis and a consecutive inflammatory response. In contrast, apoptosis proceeds in an orderly manner following a cellular suicide program involving active gene expression in response to physiological signals or types of stress. Morphologically, apoptosis is characterized by chromatin condensation and aggregation, cellular and nuclear shrinkage and formation of apoptotic bodies. It is usually not accompanied by an inflammatory response. PrP106-126 has recently been shown to cause apoptosis of rat hippocampal neurons (Forloni et al., 1993). However, apoptotic cell death has not convincingly been shown to occur in prion diseases in vivo.

The use of ISEL has greatly facilitated the recognition of apoptotic cells in tissue sections (Gavrieli et al., 1992; Gold et al., 1993). This technique was used to identify apoptotic cells in mice infected with the 79A strain of scrapie, a well-defined experimental scrapie model (Bruce et al., 1991, 1993, 1994; Fraser, 1993). To evaluate the pattern and time course of cell death, various brain regions were analyzed at different points of time in the course of the disease. Since cell loss in scrapie-infected rodents often is quite striking in the photoreceptor layer of the retina (Hogan et al., 1981, 1983; Kozlowski et al., 1982; Buyukmihci et al., 1987a, 1987b), the eyes were included in this study. Electron microscopy was used to independently assess apoptotic cell death.

Cerebellum

ISEL revealed nuclei containing fragmented DNA in the granule cell layer in scrapie-infected mice from day 120 onwards. No labeling was found in control mice at any point of time and in scrapie-infected mice up to day 90 with the exception of one scrapie-infected animal showing one labeled nucleus in ten high power visual fields 90 days after inoculation. The number of labeled cells increased from day 120 to 150 and was highest in terminally ill mice (day 166) (Fig. 1C). Labeled nuclei often appeared shrunken and showed most intense staining at the periphery of the nucleus. Corresponding to the in situ end-labeling assay, small, dark, round, occasionally fragmented nuclei with eosino-philic cytoplasm were found in sections from scrapie-infected mice stained with haematoxylin and eosin from day 120 onwards.

Electron microscopic investigation of specimens from the cerebellum of terminally ill mice identified several cells which showed homogeneously condensed chromatin, dark cytoplasm, membrane blebbing and, occasionally, nuclear fragmentation (Fig. 1D), all of which are morphological changes characteristic of apoptosis (Kerr et al., 1972; Searle et al., 1982; Sloviter et al., 1993).

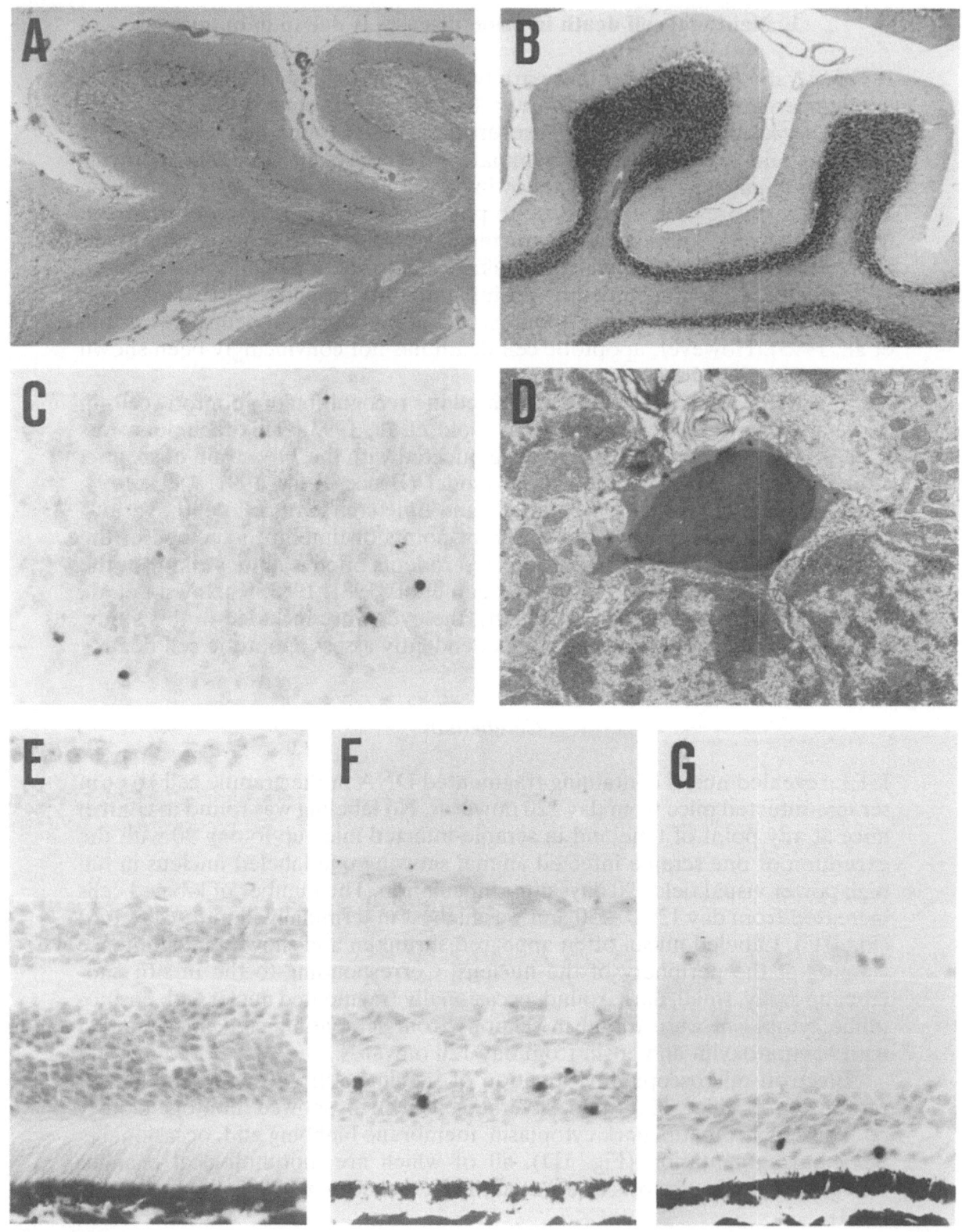

Fig. 1

Fig. 1. Neuronal cell death in prion disease. **A** Cerebellum of a patient who died from CJD. The granule cell layer is almost completely destroyed, whereas the Purkinje cells appear unaffected. Hematoxylin and eosin. **B** For comparison, this cerebellar section of an age-matched control shows normal density of granule cells. Hematoxylin and eosin. **C** In situ end-labeling of the cerebellum of a mouse terminally ill with scrapie shows several labeled nuclei. × 420. **D** Electron micrograph of an apoptotic cell in the cerebellar granule cell layer of a mouse terminally ill with scrapie. Note dense and homogeneous aggregation of nuclear chromatin, increased density of cytoplasm and membrane blebbing. Two adjacent normal granule cell nuclei are seen at the bottom. × 11660. **E–G** In situ end-labeling in the retina of scrapie-infected mice [control (E), 120 days (F) and 166 days (G) after infection] identifies apoptotic cells almost exclusively in the outer nuclear layer. The highest absolute number of labeled cells is observed at 120 days after infection. 166 days after inoculation only a minor fraction of nuclei is conserved among which there is a relatively high number of labeled nuclei. Note occasional labeling of a nucleus in the inner nuclear layer at day 166 (F). × 480.

Methods: Six-week-old female C57Bl mice were inoculated intracerebrally with 30 μl of a 10% (w/v) clarified suspension of brain in MEM [Minimal Essential Medium (Gibco)] from mice terminally affected with the 79A strain of scrapie or from age-matched uninfected C57Bl mice. Groups of 4–5 scrapie-infected mice and 2–3 controls were killed at 30 days post-injection and at 30-day intervals to 150 days, when first clinical signs of scrapie became evident, and groups of 4 terminally ill and 2 control mice were sacrificed at 166 days. Animals were perfused intracardially under nembutal anesthesia (100 mg/kg body weight) with 30 ml of 0.1% heparin in 0.9% NaCl followed by 150 ml 4% paraformaldehyde and 0.9% NaCl prepared in 0.1 M sodium-phosphate buffer (pH 7.4). The heads were removed and kept at 4°C in 4% paraformaldehyde for at least 9 days before brains and eyes were dissected, embedded in paraffin and 2.5 μm sections were cut.

For electron microscopy specimens from the cerebellum of terminally ill and control mice were used. After perfusion and fixation with 4% buffered paraformaldehyde, tissue was treated with formic acid (98–100%) for 45 minutes, kept in 3% buffered glutaraldehyde for 2 days, post-fixed in 1% osmium tetroxide, dehydrated and embedded in araldite. Semithin sections were stained with toluidine blue and ultrathin sections from selected areas were contrasted with lead citrate and examined using a Zeiss EM 10 electron microscope.

In situ end-labeling assay: After deparaffinization, sections were subjected to proteolytic pretreatment with proteinase K (Sigma) at a working dilution of 50 μg/ml in Tris-buffered saline (TBS) supplemented with 2 mM $CaCl_2$ for 20 minutes at 37°C. The slides were rinsed in TBS and then incubated under cover slips with 60 μl of the labeling mix [50 U/ml terminal deoxynucleotidyl transferase (Pharmacia), 4 μl/ml Dig-DNA labeling mix 10 × conc. (Boehringer) and 1 mM $CoCl_2$ in reaction buffer for terminal transferase (Boehringer) containing 0.2 M potassium cacodylate, 25 mM Tris-HCl and 0.25 mg/ml bovine serum albumine, pH 6.6] for 60 minutes at 37°C. After rinsing in TBS, sections were blocked with 10% fetal calf serum (FCS) (Seromed). Sections were then treated for 60 minutes with an alkaline phosphatase-labeled anti-digoxigenin antibody Fab-fragment (Boehringer) at a dilution of 1:300 in 10% FCS. After extensive washing in TBS, the color reaction was visualized by incubating the sections in reaction buffer (Tris 100 mM, NaCl 100 mM, $MgCl_2$ 50 mM, pH 9.5) containing 450 μg/ml 4-nitroblue tetrazolium salt and 175 μg/ml 5-bromo-4-chloro-3-indolyl phosphate (Boehringer) for 10 minutes. The reaction was stopped in TBS, sections were counterstained with nuclear fast red (Merck), rinsed in distilled water and mounted in Aquamount (BDH laboratory supplies).

Other regions of the brain

In the rest of the brain labeling was less obvious. As described previously (Migheli et al., 1994), labeled nuclei were observed occasionally in the ependyma and in a subependymal location along the lateral ventricles both in control and scrapie-infected mice. In scrapie-infected mice at days 150 and 166 a small number of unequivocally labeled nuclei were observed in the basal ganglia, in the granule cell layer of the olfactory bulb and in the cerebral cortex.

Retina

Sections from animals at 30, 60 and 90 days after inoculation showed no pathological changes of the retina. When compared to controls, eyes removed from animals 120 days after scrapie infection showed a dissolution of rod inner and outer segment regions (Fig. 1E, F). This process was accompanied by the presence of cells in this layer, which were identified as macrophages by electron microscopy. In two out of four animals the outer nuclear layer appeared considerably thinned. No changes were detectable in the inner nuclear layer and the ganglion cell layer. At 150 and 166 days post infection (Fig. 1G), the outer nuclear layer had degenerated to only a single cell layer in thickness with slight variations of two to three cells in some regions and a total loss of photoreceptor cells in other regions. The photoreceptor inner and outer segments had almost completely vanished. These changes were similar to those described previously (Hogan et al., 1981).

The ISEL assay showed that labeled nuclei were practically absent from retinae of control mice of all ages (Fig. 1E). Labeled nuclei were virtually absent from scrapie-infected mice up to day 90. At 120 days after infection several labeled nuclei were found scattered in the outer nuclear layer (Fig. 1F). Some labeled nuclei appeared shrunken, and labeling was often most intense at the periphery of the nucleus. At 150 and 166 days after infection solitary labeled nuclei were found in the remainder of the outer nuclear layer (Fig. 1G). However, considering the extreme loss of cells in this time interval, the percentage of apoptotic cells to the overall number of remaining cells may even have increased. Additionally, occasional labeling of single nuclei in the inner nuclear layer was observed, possibly representing transsynaptic degeneration.

Pattern and time course of apoptotic cell death in scrapie

Judging from previous studies there are indirect indications of apoptosis occurring in prion disease in vivo. Using laser scanning microscopic analysis of a single cell gel assay, increased DNA damage was found in three scrapie-infected sheep brains as compared to controls, while no attempt was made to control for necrosis, autolysis or age (Fairbairn et al., 1994). Hogan et al. (1981) gave a detailed morphological description of retinal degeneration and showed an electron micrograph of a photoreceptor cell which nowadays would be interpreted as apoptosis. Additionally, the almost complete absence of an inflammatory response in the pathology of spongiform encephalopathies (Eikelenboom et al., 1991; Williams et al., 1994) suggests that nerve cell loss may be due to apoptosis. On the other hand, in a large study investigating apoptosis in the nervous system using the ISEL technique, apoptosis was not identified in terminally ill scrapie-infected mice (Migheli et al., 1994); however, the authors give no details concerning the strains of mice and scrapie used and the brain regions investigated. We used the ISEL technique, which has recently been employed by a great number of investigators, to identify apoptotic cells in histological sections. It appears, however, that the results obtained with this technique are not absolutely specific; ISEL is based on the incorporation of labeled nucleotides in fragmented DNA by the enzyme terminal transferase which is considered highly characteristic of apoptosis. DNA fragmentation is also observed in necrosis or autolysis, but in these instances seems to occur at random, in contrast to the specific internucleosomal cleavage which is characteristic of apoptosis (Wyllie et al., 1984). Therefore, DNA laddering in gel electrophoresis, which shows the different size of oligonucleosomal fragments, is often used to distinguish apoptosis from mechanisms leading to random DNA cleavage. In cases where only a small percentage of cells undergoes apoptosis at any given moment, as expected in our experimental system, DNA laddering would not be applicable and hence was not attempted. Therefore classical morphological criteria were employed (Kerr et al., 1972; Searle et al., 1982) to validate data obtained by ISEL. These included chromatin condensation, nuclear fragmentation, eosinophilic and electron microscopically dark cytoplasm and membrane blebbing. Since all of these were present, our results convincingly demonstrate the occurrence of neuronal apoptosis in an experimental in vivo scrapie system using two different techniques.

C57Bl6 mice infected with the 79A strain of scrapie are known to show diffuse widespread accumulation of prion protein (Bruce et al., 1994) without amyloid plaques (Bruce et al., 1976; McBride et al., 1988) as well as moderate vacuolation affecting most gray matter areas including the molecular layer of the cerebellum to a fairly similar degree. Other strains of scrapie such as ME7 show comparatively little spongiform change in the cerebellum (Bruce et al., 1991). We found the most obvious labeling of apoptotic cells in the granular cell layer of the cerebellum and in the outer nuclear layer of the retina. These structures contain a high number and density of morphologically uniform neuronal cells and therefore are ideally suited for detection of a small percentage of apoptotic cells. However, other regions similar in this respect, such as the granule cell layer of the olfactory bulb or hippocampus, showed few or no labeled nuclei, indicating a different degree of cell death in these structures. It is interesting to note, however, that the highest number and density of apoptotic cell death was observed exactly in those areas that did not show spongiform changes, i.e. the retina and the granule cell layer of the cerebellum. It will be interesting to see if other experimental scrapie models such as the ME7 strain of scrapie in LM mice, which is associated with different pathological features (Scott and Fraser, 1984), show a different pattern of cell death affecting a different population of neurons. This should help establish more detailed lesion profiles and elucidate the correlation between prion protein accumulation, spongiform change and neuronal cell death.

In the retina, where the most obvious loss of cells was found, our data suggest that apoptotic cell death peaks at some point between days 90 and 150, most likely close to day 120, in this experimental system. This coincides with the time of maximal cell loss. ISEL offers the opportunity to detect this interesting early phase in the progression of the disease before massive cell loss and obvious clinical signs of scrapie are evident. The small percentage of labeled cells that were found even in areas of massive ongoing cell death is in no way surprising since apoptosis has been shown to be completed within hours (Bursch et al., 1990) and cells can be expected to become apoptotic asynchronously. This indicates that it will be difficult to demonstrate apoptotic cells in conditions where cell loss is less rapid. The demonstration of apoptosis in an animal model of scrapie therefore is of considerable importance both for the pathophysiology of spongiform encephalopathies and the validation of cell culture experiments using neurotoxic prion protein fragments as well as in the broader context of understanding the mechanisms of neurodegenerative diseases.

Since many common neurodegenerative diseases such as Alzheimer's disease, Parkinson's disease and amyotrophic lateral sclerosis are characterized by a gradual loss of neurons without obvious inflammatory response it has been hypothesized that cell death in these disorders is due to apoptosis (Thompson, 1995). However, there is only limited sound data available on apoptosis in the adult central nervous system (Sloviter et al., 1993; Su et al., 1994; Lassmann et al., 1995). Our data show that apoptosis of neurones can be demonstrated in the adult nervous system in a well characterized animal model of prion disease, which shows massive and highly predictable cell death.

2. Is PrP^C necessary for normal synaptic function?

PrP has been found in all mammals examined, has a high turnover rate (Caughey et al., 1989) and is widely expressed in early embryogenesis (Manson et al., 1992) as well as postnatally (Kretzschmar et al., 1986; Lieberburg, 1987; Moser et al., 1995). It is anchored to the neuronal membrane surface by glycosylphosphatidylinositol, suggesting a role in cell signalling or adhesion (Borchelt et al., 1993). Yet, PrP gene-ablated mice (PrP$^{0/0}$) appear developmentally and behaviorally normal (Büeler et al., 1992). These mice, however, have recently been shown to have altered synaptic transmission, giving rise to the hypothesis that PrP^C is necessary for normal synaptic function (Collinge et al., 1994). Intracellular single-electrode voltage clamp measurements in hippocampal CA1 neurons of PrP$^{0/0}$ mice showed slower inhibitory postsynaptic currents (IPSCs) compared to control animals. Depolarized reversal potentials of IPSCs suggested a postsynaptic function of PrP probably due to abnormal Cl$^-$ gradients or altered selectivity of the GABA-gated ion channels (Collinge et al., 1994). The conclusion from these experiments was that the expression of PrP^C is necessary for normal synaptic function and that neuropathological changes in prion disease may be due to the lack of functionally intact prion protein ("lack of function hypothesis").

Since the cerebellum is severely affected in most cases of human prion disease (Fig. 1A) as well as in many experimental models of scrapie, we investigated the synaptic role of prion protein in cerebellar Purkinje cells (Herms et al., 1995). Patch clamp experiments were performed in thin slice preparations to compare synaptic properties of Purkinje cells in wild type and $PrP^{0/0}$ animals (Fig. 2).

In the absence of external stimuli, IPSCs occurred in a random fashion by spontaneous activation of GABAergic synapses (Vincent et al., 1992; Llano et al., 1991). These currents were reversibly blocked by 10 µM bicuculline, a blocker of $GABA_A$ receptors. In $PrP^{0/0}$ animals, we observed spontaneous inhibitory currents in all cells investigated (n = 23) (Fig. 2A). The relative frequency of inhibitory currents depended significantly on the recorded cell but was usually found to be within 1 and 5 events per second. IPSC amplitudes in $PrP^{0/0}$ animals ranged from 20 pA to 5 nA with an average amplitude of 300 pA. This large variation was not specific for $PrP^{0/0}$ animals, since IPSCs in control animals displayed similar properties. Most likely, this variation resulted from fluctuations in the amount of neurotransmitter released at a given presynaptic site (Vincent et al., 1992). Also, there was no significant difference in the relative frequency of IPSCs between $PrP^{0/0}$ and control animals. We conclude that the lack of PrP^C did not affect the mean number of $GABA_A$ receptors activated by a single synaptic event and did not increase or reduce the average number of spontaneously active synapses in cerebellar Purkinje cells.

The activation time (10–90%) of synaptic currents was determined by fitting a linear function to the activating current as shown in Fig. 2B. The decay constant of GABA currents was determined by fitting a single exponential function to the current decay. Usually, a single exponential function was found to satisfy the statistical requirements for an appropriate fit (p < 0.01). An average activation time of 2.4 ± 0.6 ms (10–90% max. amplitude, n = 23 cells) and an average decay time constant of 11.3 ± 2.6 ms (n = 23) for the responses in $PrP^{0/0}$ animals were found. The variation was notably large, but was repeatedly observed during recordings from Purkinje cells in $PrP^{0/0}$ and control animals. As shown in Fig. 2B, the average value and statistical variation of activation and decay time constants of $PrP^{0/0}$ mice were indistinguishable from those observed under control conditions.

To further investigate the parameters of inhibitory synaptic transmission in $PrP^{0/0}$ mice, we analysed the functional parameters of isolated $GABA_A$ receptor channels in outside-out membrane patches pulled from Purkinje cells of control and $PrP^{0/0}$ mice (Fig. 2C). The problems of uncontrolled agonist diffusion around neurons in slices were avoided by performing fast application of the agonist GABA to outside-out membrane patches. This enabled us to control the temporal profile of agonist application in the submillisecond time range (Franke et al., 1987; Barbour et al., 1994). Most importantly, channel activation and deactivation was essentially equal in control and $PrP^{0/0}$ mice. Moreover, GABA currents in $PrP^{0/0}$ mice displayed a reversal potential of 2.2 ± 1.6 mV (n = 5), which was close to the reversal potential of control animals (3.7 ± 2.7 mV, n = 4).

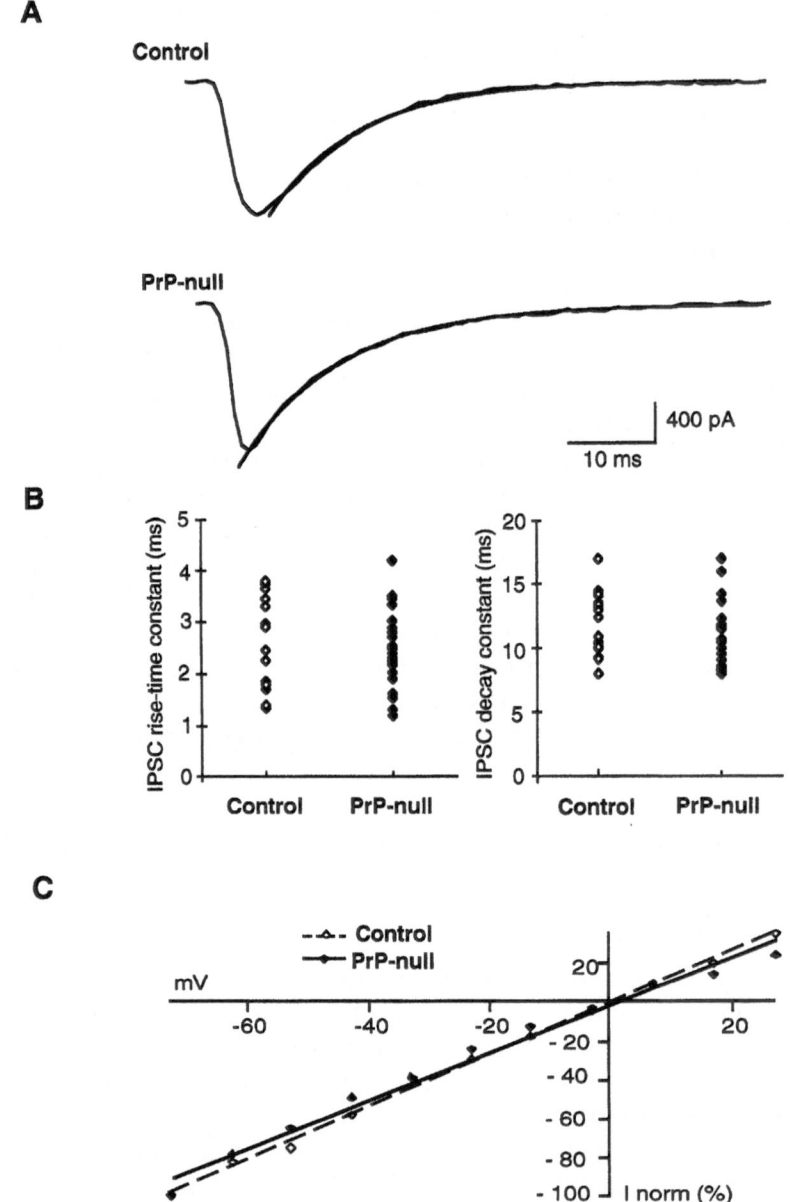

Fig. 2. Synaptic properties of PrP[0/0] cerebellar Purkinje cells. **A** Kinetic parameters of IPSCs recorded from cerebellar Purkinje cells in PrP[0/0] and control cerebellar slices. Spontaneous IPSCs recorded as inward currents at a holding potential of –63mV show similar shapes; the curve fitting used single exponentials. **B** The rise time constant did not differ significantly between PrP[0/0] cerebellar slices (2.4 ± 0.6 ms, means ± S.D., n = 23) and controls (2.7 ± 0.7 ms, n = 18), neither did the decay time constant [11.3 ± 2.6 ms in PrP[0/0] slices (n = 23) and 12 ± 2.4 ms in control slices (n = 18)]. **C** Kinetics of GABA-induced currents in outside-out patches from cerebellar Purkinje cells of PrP[0/0] and control mice. Peak current was normalized to the amplitude of –73 mV (I norm) as a function of membrane potential. The peak current-voltage relation was linear, reversing at 3.7 ± 2.7 mV in control (n = 4) and 2.2 ± 1.6 mV in PrP[0/0] mouse Purkinje cells (n = 5).

Methods: Patch clamp experiments were performed on Purkinje cells in thin slices of the cerebellum following standard procedures (Hamill et al., 1981; Edwards et al., 1989).

The pathogenesis of neurodegeneration in prion disease has long been thought to be caused by the accumulation of the pathological isoform of prion protein PrP[Sc]. Several studies investigating the neurotoxic effect of prion protein indeed showed that the exposure to either PrP[Sc] (Müller et al., 1993) or PrP peptide fragments (Forloni et al., 1993; Brown et al., 1994, 1996) leads to neuronal death in vitro. However, recent findings on hippocampal CA1 neurons (Collinge et al., 1994) questioned this hypothesis of neurodegeneration in prion disease. In hippocampal CA1 neurons of PrP[0/0] animals the kinetic parameters of IPSCs were found to be significantly slowed. A positive shift in the current voltage relation of agonist-induced GABA$_A$ receptor currents suggested that synaptic alterations might reflect modifications on the GABA$_A$ receptor level in PrP[0/0] animals (Collinge et al., 1994; Whittington et al., 1995). However, we could not reproduce any of these findings in cerebellar Purkinje cells in which synaptic transmission was found to be unimpaired in PrP[0/0] mice. There is a number of possible explanations for the different findings in hippocampal CA1 neurons and cerebellar Purkinje cells. First, PrP[C] could serve a diverse set of functions in different cell systems. The lack of PrP[C] therefore may disturb synaptic transmission in one but not the other system because the molecular machinery of synaptic transmission is different in the two cell types. This explanation has been invoked because of the remarkable heterogeneity in the time course of synaptic signals in the cerebel-

---

The two inbred lines (129/Sv and C57BL/6J) from which the PrP[0/0] mice were derived (Büeler et al., 1992) as well as the F1 cross between these were used as controls. Sagittal slices of cerebellum (150 μm) were prepared from 9 to 16 day-old mice as described (Edwards et al., 1989) and maintained at room temperature in a continuously bubbled (95% O_2, 5% CO_2) solution (125 mM NaCl, 2.5 mM KCl, 1.25 mM NaH$_2$PO$_4$, 26 mM NaHCO$_3$, 2mM CaCl$_2$, 1 mM MgCl$_2$, 25 mM glucose). After 60 min recovery, slices were placed in the recording chamber and superfused with the above solution at room temperature. A Purkinje cell was selected, using an upright microscope with a × 40 water-immersion lens (Zeiss). Electrodes were pulled from borosilicate glass capillaries and filled with a solution containing 140 mM CsCl, 10 mM HEPES, 10 mM EGTA, 1 mM CaCl$_2$, 2 mM MgCl$_2$, 4 mM Na$_2$-ATP and 0.4 mM Na$_3$-GTP (adjusted to pH 7.3 with CsOH). Single-electrode voltage-clamp recordings (Hamill et al., 1981) were performed with a patch-clamp amplifier (EPC-9, HEKA Elektronik, Germany) employing optimal series resistance compensation as recommended (Llano et al., 1991). The series resistance of Purkinje cells before compensation was usually 3–10 MOhm in control as well as in the PrP[0/0] slices. Cells with a series resistance higher than 15 MOhm before compensation were discarded from further analysis. Values of Rs and Cm were obtained from settings of the capacitance cancellation circuitry of the patch-clamp amplifier. Series resistance was set to 60–70%. Our results are derived from a data set of 34 Purkinje cells in PrP[0/0] mice and 30 Purkinje cells in control animals. All membrane potentials were corrected for the liquid junction potential (Neher, 1992).

Rapid application to outside-out patches excised from somata in PrP[0/0] and control slices was performed using a high-voltage piezoelectric crystal (Physik Instrumente, Waldbronn, Germany) to rapidly displace the agonist solution (Franke et al., 1987; Barbour et al., 1994). 1 mM GABA in bath solution was continuously perfused through an application pipette at a rate of 20–50 μl/min while the bath was continuously added at a rate of 1 ml/min with bath solution. Flows of pipette and bath solutions were arranged to be parallel to favor laminar flow of the agonist-containing solution (Barbour et al., 1994).

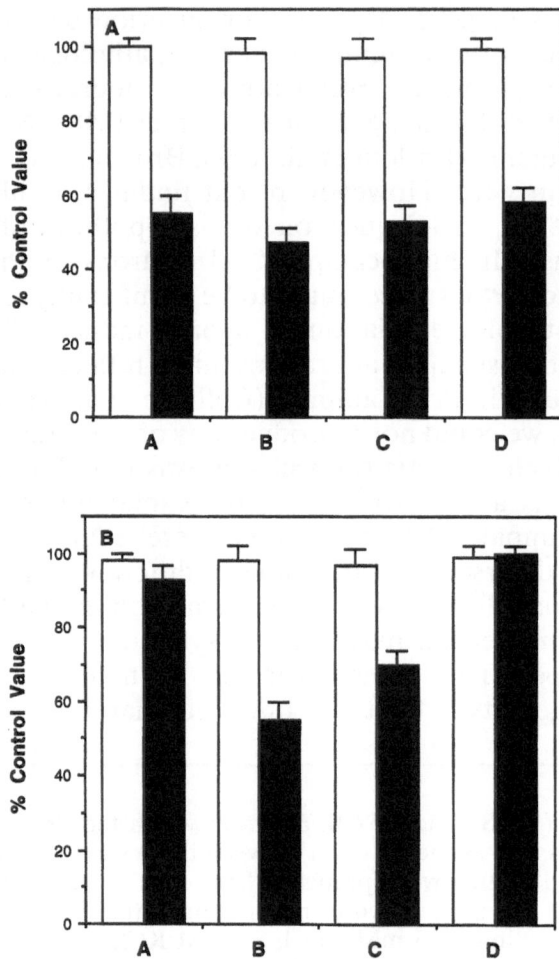

Fig. 3. Neurotoxic effects of a prion protein fragment (PrP106-126). The effect of 80 μM PrP106-126 on cerebellar cultures was determined by assessing relative survival of treated cells as compared to control cultures not treated with the peptide and expressed as a percentage of control survival. Open bars control; black bars treatment with PrP106-126. Coculture with normal microglia (B), with PrP$^{0/0}$ microglia (C), with normal astrocytes (D). This relative survival was determined by an MTT assay carried out following 10 days' treatment with the peptide. PrP106-126 treatment produced significant cell loss (Student's t test, p < 0.05). L-leucine methyl ester (LLME) treatment of cerebellar cultures did not alter cell survival but blocked the toxic effect of the peptide. Co-culturing LLME-treated cerebellar cells with astrocytes from normal mice did not alter survival of control cultures or restore the toxicity of PrP106-126. Co-culturing untreated cerebellar cells with normal microglia significantly enhanced the toxicity of PrP106-126. However, co-culturing cerebellar cells with microglia from PrP$^{0/0}$ mice did not have this effect. Shown are mean % survival ± standard error.

Methods: Cell cultures were prepared by dissociating cerebella from 6-day-old mice produced by interbreeding C57Bl/6J and 129/sv(ev) mice. The cerebella were dissociated in Hanks' media (Biochrom) containing 0.5% trypsin (Biochrom) and plated at 1–2 × 10^6 cells/cm^2 in 12 or 24 well trays coated with poly-D-lysine (50 μg/ml, Sigma). Cultures were maintained in Dulbecco's minimal essential media (Biochrom) supplemented with 10% fetal calf serum 2 mM glutamine and 1% antibiotics (penicilin, streptomycin) and kept at 37°C with 10% CO$_2$ for 10 days. Microglia were prepared from dissociating cerebral cortices (Beckman et al., 1990) of newborn normal or PrP$^{0/0}$ mice. 6–7 cortices were

lum and hippocampus (Edwards et al., 1990; Hestrin et al., 1990; Keller et al., 1991; Vincent et al., 1992; Barbour et al., 1994) and, in addition, because of the characteristic and diverse mixture of neurotransmitter receptors expressed in either cell type (Lambolez et al., 1992; Laurie et al., 1992; Nakanishi, 1992; Wisden and Seeburg, 1992; Huntley et al., 1994). Second, the discrepancies in the results could simply reflect a more flexible mechanism to compensate for the lack of PrP in Purkinje cells compared to hippocampal neurones. According to this view, the lack of PrP^C is simply compensated for by related molecular structures which would, in the course of the ontogenetic development of $PrP^{0/0}$ animals, restore the full spectrum of functional synaptic properties in Purkinje cells. At present there is no direct evidence for either one of these hypotheses. While the results presented here do not support the hypothesis of a direct role of PrP^C in synaptic transmission, they do not exclude indirect effects of PrP^C on LTP (Collinge et al., 1994).

3. In vitro models for the study nerve cell loss in prion disease

The application of a neurotoxic, synthetic prion protein (PrP) fragment has been used as a model for the investigation of neurodegeneration in prion diseases in rat hippocampal tissue cultures (Forloni et al., 1993). Using an MTT assay we demonstrated that this peptide at 80 μM is also toxic to mixed cultures from the cerebellum of 6 day-old normal mice (Fig. 3) (Brown et al., 1996). However, at 20 μM this peptide was not toxic to our cultures, with survival of 97 ± 4% of cells as compared to untreated controls. Cultures were also treated with 80 μM of a control peptide (with the amino acid sequence of

---

trypsinized in 0.5% trypsin and plated in 75 cm^2 culture flasks (Flacon). Cultures were maintained at 37°C with 10% CO_2 for 14 days until glial cultures were confluent. Microglia were isolated from these cultures by shaking cultures at 260 r.p.m. for two hours. Astrocytes were prepared from the same glia preparations after most microglia had been removed by shaking. Further trypsination of adherent cells and preplating for 30 min to remove remaining microglia gave pure astrocyte cultures. Peptides used in these experiments were PrP106-126 with the sequence KTNMKHMAGAAAAGAVVGGLG, and as a control PrP106-126 scrambled which consists of the same amino acids as PrP106-126 but in a random order NGAKALMGGHGATKVMVGAAA. Peptides were synthesized on a Milligan 9050 peptide synthesizer and added to cultures at 2-day intervals from initial plating day until the 8th day. The peptide concentration was maintained at 80 μM. Cell survival was determined on day 10. MTT (3,[4,5 dimethylthylthiazol2yl]-2,5 diphenyltetrazolium bromide, Sigma) was diluted to 200 μM in Hanks' solution (Biochrom) and added to cultures for 2 hours at 37°C. The MTT formazan product was released from cells by addition of dimethylsulphoxide (Sigma) and measured at 570 nm in an Ultrospec III spectrophotometer (Pharmacia). Relative survival in comparison to untreated controls could then be determined. L-leucine methyl ester (Sigma) was applied to cultures one day after plating, at 5 mM. Two hours after application the media was removed and fresh media with or without peptides applied. Cerebellar cells were fixed on coverslips for coculturing and microglia or astrocytes plated at 10^5 cells/cm^2 were prepared in 24 well cloning trays. Coverslips were then transferred to the cloning trays for experiments. Before MTT assays, the coverslips were removed and placed in fresh trays.

PrP106-126 in a random order). This scrambled peptide was also non-toxic ($99 \pm 3\%$ survival compared to control, n = 5).

We investigated the involvement of different cell types present in mixed cerebellar cultures in the neurotoxic mechanism of PrP106-126. Co-culturing cerebellar cells with microglia derived from normal mouse cortex leads to a slight increase in the percentage of cells destroyed but, when co-cultured with microglia from PrP$^{0/0}$ mice, there was no significant increase in cell loss (Fig. 3). To further study the role of microglia, we used a two-hour treatment with L-leucine methyl ester (LLME) which significantly reduces the number of microglia in mixed cultures (Giulian and Baker, 1986; Giulian et al., 1988) without reducing the overall number of surviving cells. However, after this treatment, 10-day administration of 80 μM PrP106-126 was no longer toxic to cerebellar cells (Fig. 3), showing that the presence of microglia is a necessary requirement for the neurotoxicity of PrP106-126.

If LLME treatment prevents the neurotoxicity of PrP106-126 by destruction of microglia, then co-culturing LLME-treated cerebellar cultures with microglia should restore the toxic effect. Co-culturing LLME-treated cells from normal mice with purified microglia from cerebral cortex of either normal or PrP$^{0/0}$ mice resulted in restoration of the toxic effect (Fig. 3). This effect is not seen when LLME-treated cerebellar cells were co-cultured with astrocytes from normal mouse cortex. Also, co-culture with normal astrocytes did not enhance the toxicity of PrP106-126 on cerebellar cultures not treated with LLME. This indicates that the neurodestructive mechanism of PrP106-126 does not involve a direct effect on astrocytes.

When cerebellar cells from PrP$^{0/0}$ mice were treated with PrP106-126 using the same method as for normal cells the peptide showed no toxic effect. After 10 days' treatment with PrP106-126 at 80 μM PrP$^{0/0}$ cerebellar cells showed a slight but significant (Student's t test, $p < 0.05$) increase in number. When cerebellar cells derived from PrP$^{0/0}$ mice were co-cultured with high numbers of microglia from normal mice, the response to PrP106-126 remained unaltered (non-toxic to PrP$^{0/0}$ cells) even when applied at very high concentrations. These findings again show that the neurotoxic effect of PrP106-126 is based on two requirements: a specific interaction with PrP^{C}-expressing neurones and the presence of microglia.

If neurones are destroyed as a result of substances released by microglia stimulated by PrP106-126, then it should be possible to block the effect of these substances in culture. A strong candidate for this effect would be an oxidative substance such as the superoxide radical or a reactive by-product of it such as nitrite, both of which are known to be released by microglia under certain conditions (Giulian and Baker, 1986; Betz Corradin et al., 1993). Anti-oxidants vitamin E and N-acetyl-cysteine both blocked the toxic effect of PrP106-126 on cultures of cerebellar neurones (data not shown) (Brown et al., 1996). This supports the notion that microglia place oxidative stress on neurones as a result of PrP106-126 treatment.

Superoxide radicals and nitric oxide both can be produced by microglia (Betz Corradin et al., 1993) and induce oxidative stress. Both are short-lived molecules that readily react to produce nitrite, which is stable in solution

(Beckman et al., 1990). We assayed the supernatants of microglial cultures for nitrite production and found a high increase in production of nitrite in solution after application of PrP106-126 for 4 days. This increase was directly proportional to the log number of microglia present. However, this effect was less clearly pronounced in PrP$^{0/0}$ microglia (Brown et al., 1996). This result provides evidence that PrP106-126 acts directly on microglia, and gives strong support for the notion that microglia induce neuronal death by oxidative stress as a result of PrP106-126 stimulation.

PrP106-126 is apparently non-toxic to LLME-treated cerebellar cells, presumably because the source of oxidative stress induced by the peptide, namely microglia, has been destroyed. However, if the effect on microglia, resulting in oxidative stress, were the only effect of PrP106-126, then one would expect PrP$^{0/0}$ mouse cerebellar cells to be destroyed in a similar way. To test this hypothesis, xanthine oxidase was used as an alternative source to produce superoxide radicals. Treatment of cerebellar cells with xanthine oxidase (Fig. 4) indicates that PrP$^{0/0}$ mouse cells are more susceptible to oxidative stress. However, PrP106-126 enhances cell death induced by xanthine oxidase only on LLME-treated cerebellar cells from normal mice. This strongly indicates that for PrP106-126 to exert a toxic effect, cells must express PrP^C.

	Mouse strain	
Treatment	Normal	PrP$^{0/0}$
Control	100 ± 4	100 ± 2
Xanthine oxidase	90 ± 5	75 ± 4
Xanthine oxidase with 80 μM PrP106-126 scrambled	92 ± 5	77 ± 2
Xanthine oxidase with 80 μM PrP106-126	65 ± 5	79 ± 3

Fig. 4. The role of oxidative stress in PrP106-126 neurotoxicity. Cultures of normal or PrP$^{0/0}$ cerebellar cells were treated with LLME to destroy microglia. The effect of 80 μM PrP106-126 on these cell cultures was then determined in the presence of a known oxygen radical producer, xanthine oxidase. A concentration of xanthine oxidase was used that had a slight effect on the survival of normal LLME-treated cerebellar cells as compared to controls. The effect of the same concentration of xanthine oxidase on PrP$^{0/0}$ cerebellar cells is significantly greater (Student's t test, $p < 0.05$), suggesting they are less resistant to oxidative stress. 80 μM PrP106-126 does not kill LLME-treated cerebellar cells (see Fig. 3). However, 80 μM PrP106-126 greatly enhanced the effect of xanthine oxidase on LLME-treated cerebellar cells from normal mice but not those from PrP$^{0/0}$ mice. The scrambled peptide applied at 80 μM in conjunction with xanthine oxidase does not have this effect. This suggests that PrP106-126 treatment greatly enhances susceptibility to oxidative stress only to those cells expressing PrP^C. – Shown are mean and standard errors

Methods: Cerebellar cultures, LLME treatment and survival assay were as described in Fig. 3. Cultures treated with xanthine oxidase were exposed to 5 μUnits of xanthine oxidase (Boehringer) in the presence of 5 μM xanthine for 24 hours. Xanthine oxidase treated cultures were also co-exposed to 80 μM PrP106-126 for the same time interval after which MTT survival assays were performed.

PrP106-126 has been considered as a model for the investigation of the pathogenesis of prion diseases (Forloni et al., 1993). However, unlike PrP[Sc], PrP106-126 has not been shown to be associated with infectivity or cause conversion of PrP[C] to PrP[Sc]. On the other hand, both PrP[Sc] and PrP106-126 are neurotoxic in cell culture (Forloni et al., 1993; Müller et al., 1993) and induce apoptosis (Forloni et al., 1993; Giese et al., 1995). In addition, our data are in favour of a specific interaction of PrP106-126 with PrP[C]-expressing neurones and may therefore represent one particular aspect of pathogenesis in prion disease.

It is clear from our data that the toxic effect which PrP106-126 exerts on neurones in cell culture is dependent on the presence of and is induced by microglia. Oxidative stress is involved in this mechanism. Microglia have also been reported to be activated in murine scrapie in vivo (Williams et al., 1994). Also, recent strong evidence has appeared (Frackowiak et al., 1992; Meda et al., 1995) suggesting that microglia are stimulated by β-amyloid to produce reactive nitrogen intermediates and tumor necrosis factor (TNF-α). Microglia, whose function has largely been regarded as scavengers and mediators of immunological reactions, thus appear to emerge as an important factor inducing neuronal death in degenerative diseases.

Conclusions

Prion research in the past was focused on the biochemistry of the infectious agent. Questions concerning cellular pathogenesis and the processes underlying neuronal cell death have only recently been addressed. It appears that nerve cell death in prion disease follows apoptotic mechanisms in vitro and in vivo. While some groups hold that neuronal apoptosis is caused by progressive loss of function of PrP[C] in disease, others argue that PrP[Sc] has toxic effects on host neurones. We have demonstrated that PrP106-126, a synthetic PrP peptide, has toxic effects on neurones expressing PrP[C] while neurones devoid of PrP[0/0] are resistant. The toxic effect of PrP106-126 requires the presence of microglia in culture, which respond by producing oxygen radicals. It will be crucial to test these cell culture findings in vivo.

Acknowledgements

The authors thank C. Weissmann, University of Zürich, for providing PrP[0/0] mice. We thank Roswitha Fischer, Wolfgang Dröse, Brigitte Maruschak and Barbara Lage for technical assistance as well as Cynthia Bunker for editing of the manuscript. This study was supported by research grants from the Bundesministerium für Bildung, Wissenschaft, Forschung und Technologie, the Deutsche Forschungsgemeinschaft (DFG), the Wilhelm-Sander-Stiftung and the Fritz-Thyssen-Stiftung.

References

Barbour B, Keller BU, Llano I, Marty A (1994) Prolonged presence of glutamate during excitatory synaptic transmission to cerebellar Purkinje cells. Neuron 12: 1331–1343

Beckman JS, Beckman TW, Chen J, Marshall PA, Freeman BA (1990) Apparent hydroxyl radical production by peroxynitrite: Implications for endothelial injury from nitric oxide and superoxide. Proc Natl Acad Sci USA 87: 1620–1624

Betz Corradin S, Mauel J, Donini SD, Quattrocchi E, Ricciardi-Castagnoli P (1993) Inducible nitric oxide synthase activity of cloned murine micro glial cells. Glia 7: 255–262

Borchelt DR, Rogers M, Stahl N, Telling G, Prusiner SB (1993) Release of the cellular prion protein from cultured cells after loss of its glycoinositol phospholipid anchor. Glycobiology 3: 319–330

Brown DR, Herms J, Kretzschmar HA (1994) Mouse cortical cells lacking cellular PrP survive in culture with a neurotoxic PrP fragment. Neuro Report 5: 2057–2060

Brown DR, Schmidt B, Kretzschmar HA (1996) Role of microglia and host prion protein in neurotoxicity of prion protein fragment. Nature 380: 345–347

Bruce ME, Dickinson AG, Fraser H (1976) Cerebral amyloidosis in scrapie in the mouse: effect of agent strain and mouse genotype. Neuropathol Appl Neurobiol 2: 471–478

Bruce ME, McConnell I, Fraser H, Dickinson AG (1991) The disease characteristics of different strains of scrapie in Sinc congenic mouse lines: implications for the nature of the agent and host control of pathogenesis. J Gen Virol 72: 595–603

Bruce ME (1993) Scrapie strain variation and mutation. Br Med Bull 49: 822–838

Bruce ME, McBride PA, Jeffrey M, Scott JR (1994) PrP in pathology and pathogenesis in scrapie-infected mice. Mol Neurobiol 8: 105–112

Buja LM, Eigenbrodt ML, Eigenbrodt EH (1993) Apoptosis and necrosis. Basic types and mechanisms of cell death. Arch Pathol Lab Med 117: 1208–1214

Bursch W, Paffe S, Putz B, Barthel G, Schulte-Hermann R (1990) Determination of the length of the histological stages of apoptosis in normal liver and in altered hepatic foci of rats. Carcinogenesis 11: 847–853

Buyukmihci NC, Goehring-Harmon F, Marsh RF (1987a) Photoreceptor degeneration during infection with various strains of the scrapie agent in hamsters. Exp Neurol 97: 201–206

Buyukmihci NC, Goehring-Harmon F, Marsh RF (1987b) Photoreceptor degeneration in experimental transmissible mink encephalopathy of hamsters. Exp Neurol 96: 727–731

Büeler H, Fischer M, Lang Y, Bluethmann H, Lipp H-P, DeArmond SJ, Prusiner SB, Aguet M, Weissmann C (1992) Normal development and behaviour of mice lacking the neuronal cell-surface PrP protein. Nature 356: 577–582

Caughey B, Race RE, Ernst D, Buchmeier MJ, Chesebro B (1989) Prion protein biosynthesis in scrapie-infected and uninfected neuroblastoma cells. J Virol 63: 175–181

Collinge J, Whittington MA, Sidle KCL, Smith CJ, Palmer MS, Clarke AR, Jefferys JGR (1994) Prion protein is necessary for normal synaptic function. Nature 370: 295–297

Edwards FA, Konnerth A, Sakmann B, Takahashi T (1989) A thin slice preparation for patch-clamp recordings from neurones of the mammalian central nervous system. Pflugers Arch 414: 600–612

Edwards FA, Konnerth A, Sakmann B (1990) Quantal analysis of inhibitory synaptic transmission in the dentate gyrus of rat hippocampal slices: a patch-clamp study. J Physiol 430: 213–249

Eikelenboom P, Rozemuller JM, Kraal G, Stam FC, McBride PA, Bruce ME, Fraser H (1991) Cerebral amyloid plaques in Alzheimer's disease but not in scrapie-affected mice are closely associated with a local inflammatory process. Virchows Arch [B] 60: 329–336

Fairbairn DW, Carnahan KG, Thwaits RN, Grigsby RV, Holyoak GR, ONeill KL (1994) Detection of apoptosis induced DNA cleavage in scrapie-infected sheep brain. FEMS Microbiol Lett 115: 341–346

Forloni G, Angeretti N, Chiesa R, Monzani E, Salmona M, Bugiani O, Tagliavini F (1993) Neurotoxicity of a prion protein fragment. Nature 362: 543–546

Frackowiak J, Wisniewski HM, Wegiel J, Merz GS, Iqbal K, Wang KC (1992) Ultrastructure of the microglia that phagocytose amyloid and the microglia that produce β-amyloid fibrils. Acta Neuropathol (Berl) 84: 225–233

Franke C, Hatt H, Dudel J (1987) Liquid filament switch for ultra-fast exchanges of solutions at excised patches of synaptic membrane of crayfish muscle. Neurosci Lett 77: 199–204

Fraser H (1993) Diversity in the neuropathology of scrapie-like diseases in animals. Br Med Bull 49: 792–809

Gavrieli Y, Sherman Y, Ben-Sasson SA (1992) Identification of programmed cell death in situ via specific labeling of nuclear DNA fragmentation. J Cell Biol 119: 493–501

Giese A, Groschup MH, Hess B, Kretzschmar HA (1995) Neuronal cell death in scrapie-infected mice is due to apoptosis. Brain Pathol 5: 213–221

Giulian D, Young DG, Woodward J, Brown DC, Lachman LB (1988) Interleukin-1 is an astroglial growth factor in the developing brain. J Neurosci 8: 709–714

Giulian D, Baker TJ (1986) Characterization of ameboid microglia isolated from developing mammalian brain. J Neurosci 6: 2163–2178

Gold R, Schmied M, Rothe G, Zischler H, Breitschopf H, Wekerle H, Lassmann H (1993) Detection of DNA fragmentation in apoptosis: application of in situ nick translation to cell culture systems and tissue sections. J Histochem Cytochem 41: 1023–1030

Hamill O, Marty A, Neher E, Sakmann B, Sigworth FJ (1981) Improved patch-clamp techniques for high resolution current recording from cells and cell-free membrane patches. Pflugers Arch 391: 85–100

Herms JW, Kretzschmar HA, Titz S, Keller BU (1995) Patch-clamp analysis of synaptic transmission to cerebellar Purkinje cells of prion protein knockout mice. Eur J Neurosci 7: 2508–2512

Hestrin S, Nicoll RA, Perkel DJ, Sah P (1990) Analysis of excitatory synaptic action in pyramidal cells using whole-cell recording from rat hippocampal slices. J Physiol 422: 203–225

Hogan RN, Baringer JR, Prusiner SB (1981) Progressive retinal degeneration in scrapie-infected hamsters: a light and electron microscopical analysis. Lab Invest 44: 34–42

Hogan RN, Kingsbury DT, Baringer JR, Prusiner SB (1983) Retinal degeneration in experimental Creutzfeldt-Jakob disease. Lab Invest 49: 708–715

Huntley CW, Vickers JC, Morrison JH (1994) Cellular and synaptic localization of NMDA and non-NMDA receptor subunits in neocortex: organizational features related to cortical circuitry, function and disease. TINS 17: 536

Jakob A (1921) Über eigenartige Erkrankungen des Zentralnervensystems mit bemerkenswertem anatomischem Befunde (spastische Pseudosklerose-Encephalomyelopathie mit disseminierten Degenerationsherden). Dtsch Z Nervenheilkd 70: 132–146

Jakob A (1923) Spastische Pseudosklerose. In: Jakob A (ed) Die extrapyramidalen Erkrankungen. Springer, Berlin, pp 215–245

Jeffrey M, Halliday WG, Goodsir CM (1992) A morphometric and immunohistochemical study of the vestibular nuclear complex in bovine spongiform encephalopathy. Acta Neuropathol (Berl) 84: 651–657

Jeffrey M, Fraser JR, Halliday WG, Fowler N, Goodsir CM, Brown DA (1995) Early unsuspected neuron and axon terminal loss in scrapie-infected mice revealed by morphometry and immunocytochemistry. Neuropathol Appl Neurobiol 21: 41–49

Keller BU, Konnerth A, Yaari Y (1991) Patch clamp analysis of excitatory synaptic currents in granule cells of rat hippocampus. J Physiol 435: 275–293

Kerr JFR, Wyllie AH, Currie AR (1972) Apoptosis: a basic biological phenomenon with wide-ranging implications in tissue kinetics. Br J Cancer 26: 239–257

Kozlowski PB, Moretz RC, Carp RI, Wisniewski HM (1982) Retinal damage in scrapie mice. Acta Neuropathol (Berl) 56: 9–12

Kretzschmar HA, Prusiner SB, Stowring LE, DeArmond SJ (1986) Scrapie prion proteins are synthesized in neurons. Am J Pathol 122: 1–5

Lambolez B, Audinat E, Bochet P, Crepel F, Rossier J (1992) AMPA receptor subunits expressed by single Purkinje cells. Neuron 9: 247–258

Lassmann H, Bancher C, Breitschopf H, Wegiel J, Bobinski M, Jellinger K, Wisniewski HM (1995) Cell death in Alzheimer's disease evaluated by DNA fragmentation in situ. Acta Neuropathol (Berl) 89: 35–41

Laurie DJ, Seeburg PH, Wisden W (1992) The distribution of 13 GABA$_A$ receptor subunit mRNAs in the rat brain. II. Olfactory bulb and cerebellum. J Neurosci 12: 1063–1076

Lieberburg I (1987) Developmental expression and regional distribution of the scrapieassociated protein mRNA in the rat central nervous system. Brain Res 417: 363–366

Llano I, Marty A, Armstrong CM, Konnerth A (1991) Synaptic and agonist-induced currents of Purkinje cells in rat cerebellar slices. J Physiol 434: 183–213

Manson J, West JD, Thomson V, McBride P, Kaufman MH, Hope J (1992) The prion protein gene: a role in mouse embryogenesis? Development 115: 117–122

Masters CL, Richardson EP Jr (1978) Subacute spongiform encephalopathy (Creutzfeldt-Jakob disease). The nature and progression of spongiform change. Brain 101: 333–344

McBride PA, Bruce ME, Fraser H (1988) Immunostaining of scrapie cerebral amyloid plaques with antisera raised to scrapie-associated fibrils (SAF). Neuropathol Appl Neurobiol 14: 325–336

Meda L, Cassatella MA, Szendrei GI, Otvos L, Baron P, Villalba M, Ferrari D, Rossi F (1995) Activation of microglial cells by β-amyloid protein and interferon-gamma. Nature 374: 647–650

Migheli A, Cavalla P, Marino S, Schiffer D (1994) A study of apoptosis in normal and pathologic nervous tissue after in situ end-labeling of DNA strand breaks. J Neuropathol Exp Neurol 53: 606–616

Moser M, Colello RJ, Pott U, Oesch B (1995) Developmental expression of the prion protein gene in glial cells. Neuron 14: 509–517

Müller WEG, Ushijima H, Schroder HC, Forrest JMS, Schatton WFH, Rytik PG, Heffnerlauc M (1993) Cytoprotective effect of NMDA receptor antagonists on prion protein (Prion^{Sc})-induced toxicity in rat cortical cell cultures. Eur J Pharmacol 246: 261–267

Nakanishi S (1992) Molecular diversity of glutamate receptors and implications for brain function. Science 258: 597–603

Neher E (1992) Correction for liquid junction potentials in patch clamp experiments. Meth Enzymol 207: 123–131

Prusiner SB (1982) Novel proteinaceous infectious particles cause scrapie. Science 216: 136–144

Prusiner SB (1991) Molecular biology of prion diseases. Science 252: 1515–1522

Prusiner SB (1993) Genetic and infectious prion diseases. Arch Neurol 50: 1129–1153

Scott JR, Fraser H (1984) Degenerative hippocampal pathology in mice infected with scrapie. Acta Neuropathol (Berl) 65: 62–68

Searle J, Kerr JFR, Bishop CJ (1982) Necrosis and apoptosis: distinct modes of cell death with fundamentally different significance. Path Ann 17: 229–259

Sloviter RS, Dean E, Neubort S (1993) Electron microscopic analysis of adrenalectomy-induced hippocampal granule cell degeneration in the rat: apoptosis in the adult central nervous system. J Comp Neurol 330: 337–351

Spielmeyer W (1922) Die histopathologische Forschung in der Psychiatrie. Klin Wochenschr 1: 1817–1819

Su JH, Anderson AJ, Cummings BJ, Cotman CW (1994) Immunohistochemical evidence for apoptosis in Alzheimer's disease. Neuroreport 5: 2529–2533

Thompson CB (1995) Apoptosis in the pathogenesis and treatment of disease. Science 267: 1456–1462

Vincent P, Armstrong CM, Marty A (1992) Inhibitory synaptic currents in rat cerebellar
    Purkinje cells: modulation by postsynaptic depolarization. J Physiol 456: 453–471
Whittington MA, Sidle KCL, Gowland I, Meads J, Hill AF, Palmer MS, Jefferys JGR,
    Collinge J (1995) Rescue of neurophysiological phenotype seen in PrP null mice by
    transgene encoding human prion protein. Nat Genet 9: 197–207
Williams AE, Lawson LJ, Perry VH, Fraser H (1994) Characterization of the microglial
    response in murine scrapie. Neuropathol Appl Neurobiol 20: 47–55
Wisden W, Seeburg PH (1992) GABA$_A$ receptor channels: from subunits to functional
    entities. Curr Opin Neurobiol 2: 263–269
Wyllie AH, Morris RG, Smith AL, Dunlop D (1984) Chromatin cleavage in apoptosis:
    association with condensed chromatin morphology and dependence on macromolecu-
    lar synthesis. J Pathol 142: 67–77

Authors' address: Prof. Dr. H. A. Ketzschmar, Institute of Neuropathology, University of Göttingen, Robert-Koch-Strasse 40, D-37075 Göttingen, Federal Republic of
Germany.

Subject Index

SpringerNeurology

Wilfried Kuhn, Peter Kraus, Horst Przuntek (eds.)

Deprenyl – Past and Future

1996. 16 figures. IX, 112 pages. Soft cover DM 104,–, öS 728,–
Only available for subscribers to "Journal of Neural Transmission"
ISBN 3-211-82891-5
Journal of Neural Transmission, Supplement 48

Cloth DM 130,–, öS 910,–
(Special edition of "Journal of Neural Transmission", Supplement 48)
ISBN 3-211-82948-2

The clinical effect of L-Deprenyl was originally explained on the basis of irreversible and selective MAO-B inhibition and subsequent enhancement of dopaminergic neurotransmission. In recent years new experimental data have challenged this concept. In vitro and in vivo studies are suggesting that L-Deprenyl may have neuroprotective and/or neuroregenerative properties, too. Furthermore, controversial data of recently finished long-term clinical studies have brought forward an new discussion both on the clinical impact and the possible mode of action of L-Deprenyl in Parkinson's Disease and various other neurological and psychiatric disorders. This volume provides a forum for intensive discussions on the biochemical, pharmacological and clinical aspects of Parkinson's Disease.

Kurt A. Jellinger, Manfred Windisch (eds.)

New Trends in the Diagnosis and Therapy of Non-Alzheimer's Dementia

1996. 61 partly coloured figures. VIII, 288 pages. Soft cover DM 190,–, öS 1330,–
Reduced price for subscribers to "Journal of Neural Transmission":
Soft cover DM 171,–, öS 1197,–. ISBN 3-211-82823-0
Journal of Neural Transmission, Supplement 47

This volume gives an overview of the present state of art on the classification, neuropathology, clinical presentation, neuropsychology, diagnosis, neuroimaging and therapeutic possibilities in non-Alzheimer's dementias, an increasingly important group of CNS diseases, which account for 7 to 30% of dementing disorders in adults and aged subjects, and thus, represent the second most frequent cause of dementia after Alzheimer's disease. The monograph provides the newest information for neurologists, psychiatrists, dementia research workers, dementia clinicians, neuropathologists, neurobiologists, and practicing physicians.

SpringerWienNewYork

P.O.Box 89, A-1201 Wien • New York, NY 10010, 175 Fifth Avenue
Heidelberger Platz 3, D-14197 Berlin • Tokyo 113, 3-13, Hongo 3-chome, Bunkyo-ku